T0360578

Knots, Braids and Möbius Strips

Particle Physics and the Geometry of Elementarity:
An Alternative View

Series on Knots and Everything — Vol. 55

Knots, Braids and Möbius Strips

Particle Physics and the Geometry of Elementarity: An Alternative View

Jack Avrin

NEW JERSEY • LONDON • SINGAPORE • BEIJING • SHANGHAI • HONG KONG • TAIPEI • CHENNAI

Published by

World Scientific Publishing Co. Pte. Ltd.
5 Toh Tuck Link, Singapore 596224
USA office: 27 Warren Street, Suite 401-402, Hackensack, NJ 07601
UK office: 57 Shelton Street, Covent Garden, London WC2H 9HE

Library of Congress Cataloging-in-Publication Data
Avrin, Jack, author.
Knots, braids and Möbius strips : particle physics and the geometry of elementarity : an alternative view / by Jack Avrin.
pages cm. -- (Series on knots and everything ; vol. 55)
Includes bibliographical references.
ISBN 978-981-4616-00-3 (alk. paper)
1. Particles (Nuclear physics)--Mathematics. 2. Geometry, Algebraic. 3. Knot theory. 4. Topology. I. Title.
QC793.3.G46A97 2015
539.7'2--dc23

2014028714

British Library Cataloguing-in-Publication Data
A catalogue record for this book is available from the British Library.

Typeset by Stallion Press
Email: enquiries@stallionpress.com

Printed in Singapore

To Charlene;
Now and Forever

Acknowledgments

It is impossible to thank all those from whose books and papers I learned something about the subject matter, the interpretation, variation and extrapolation of which make up the substance of this book. However, it may not be inappropriate to list in one place those whose contributions are cited directly in the book and who (unknowingly) provided me with support and encouragement at various points in this journey I undertook, in all innocence, into the mysteries and perils of particulate elementarity.[1] They include the following:

A. F. Möbius, E. Cartan, W. H. Thompson (Lord Kelvin), W. K. Clifford, A. Einstein, N. Bohr, P. A. M. Dirac, E. Fermi & C. Y. Yang, Y. Nambu, S. Sakata, J. A. Wheeler, U. Enz, S. Barr, L. Ricca & M. A. Berger, L. Fadeev & A. J. Niemi, W. A. Perkins, A. T. Filippov, L. H. Kauffman, A. Sossinsky, D. Stone, S. Sternberg, C. J. Isham, E. Kreysig, E. Flapan, S. Weinberg, W. P. Rolnick, K. Gottfried & V. Weisskopf, M. Kaku, P. D. V. Collins & M. D. Martin & E. J. Squires, F. Mandl & G. Shaw, P. E. Coghlin & J. E. Dodd, P. J. Peebles, A. Zee, J. Baez, L. O'Raifertaigh, M. Rees, J. D. Barrow & F. J. Tipler, L. A. Vilenkin, B. Greene, E. Witten, M. Atiyah, A. Ashtekar and J. Stachel.

Also, special thanks must go to Professor Louis H. Kauffman, UIC, Editor-in-Chief of the Journal of Knot Theory and its

[1]I'm afraid I may have invented a word but, to me, it's so expressive that I shall continue to use it!

Ramifications (JKTR) and of the World Scientific Series on Knots and Everything for his unflagging support over the last decade or so for my developing notions of the connection between Möbius strips and elementary particle physics and for his suggestion that I write my intended book as part of that prestigious series. Also very helpful were the comments by Professor Emeritus Lewis Licht, UIC, calling my attention to misleading or erroneous items in a paper to be used as background for the differential geometric part of the book.

Also, to Dr. William F. Avrin, my sincere appreciation for his portrayal of a concatenated trefoil knot on the cover of this book (as well as in the body of the manuscript along with related displays); as will be seen, that way of modeling a Möbius strip is important enough to justify its enigmatic appearance on the cover. And, finally, last but not least, it is really impossible to adequately express how I feel about Charlene N. Avrin for a lifetime of physical, emotional, intellectual, philosophical and gastronomical support. Plus some superhuman patience and understanding during the evolution of this book.

Contents

Preface

We have found a strange footprint on the shores of the unknown. We have devised profound theories, one after another to account for its origin. At last, we have succeeded in reconstructing the creature that made the footprint. And lo! It is our own. (Sir Arthur Eddington, "Space, Time and Gravitation").

I'm glad you're reading this Preface; my main purpose here is to express something like a point of view that underlies the substance of the book. I realize that in so doing I shall be navigating perilously close to some deep and important matters of a philosophical as well a technical nature that I would prefer to avoid at this point. If I fail, perhaps you can let me know; all comments are welcome. First, about that subtitle: although I'm not really sure that elementarity is a real word (an attention getter, maybe?) I feel so positive about the connection between the elementary particles and geometry that I once wrote a paper entitled *A Geometrical Model of the Elementary Particles*. I don't think I would say anything quite like that now but let me at least sketch out here why I feel that way about the particle/geometry connection; the rest of the book should convince you, I trust.

I'll start right out with some heavy stuff. As creatures go, the human species is anomalous. I'm sure you'll agree. Of course, we have the same needs and urges as the rest of them. And the same limitations of life and death apply to all of us, even the laws of thermodynamics. Apparently, however, we are unique in that somehow we have evolved (or, if you prefer, been endowed with) this additional capacity, be it blessing or curse, the urge to create, to wonder, even to speculate and hypothesize, that we share with none of the other denizens of our planet. It's a part of our nature that takes us far and wide and, finally, up against the ultimate questions: what it's all about — the meaning of life, even of existence — and, if there is such, what is the ultimate reality? I've been there, confronted by that frustrating impasse, and you probably have too. And maybe you've come to the same conclusion I have: that we'll probably never know the answers. Being the creatures that we are we may not even ever know whether there *are* answers. Nevertheless, apparently also, being the creatures that we are, we'll never stop trying. So, if it's not too pretentious, let me tell you what I think we *can* do, in fact what it is that we really have been doing and probably will always do. It may seem like a diversion but please bear with me.

When I was a young schoolboy, I was a prolific reader. The branch library was within walking distance and I spent a lot of time there, reading all kinds of stuff, much of which I didn't fully understand. Two books in that category come to mind: *Gulliver's Travels* by Jonathon **Swift** and one by Sir Arthur **Eddington**, renowned in the early 20th century as physicist–astronomer and a philosopher-popularizer of science (see above). I loved the Gulliver book for the fantasy of coping in worlds populated by beings much larger or much smaller than oneself. It wasn't until I was in high school that I read the book again and realized that what Swift had created was an *allegory* (in fact, a devastating commentary) of the way things were in his part of the real world. (The teacher let me prattle on about it for two hours!)

I remembered little of Eddington but two (also allegorical!) items stuck with me: one was something about an icthyologist who repeatedly cast his net into the sea only to conclude that it contained no fish smaller than a certain size (that's right; you got it!). The other I remembered (erroneously!) as something about a castaway finding a footprint on the sand and finally realizing that he had made it himself. Recently, I found the reference, fittingly, I think, in a fascinating book on cosmology by Alex **Vilenkin** [1] and it wasn't quite what I had remembered but there it is: right at the top of this Preface! Following up on that line of thought, we see that the influence and limitations of human perception, indeed, of the very nature of human intellect, preconditions the kind of description and analysis we can apply to nature as we experience it. In other words, innate point of view, or, in more currently colloquial terms, "where you're coming from" is all important.

So what does all this have to do with what we can say about "Ultimate Reality"? Well, when we think about it, for us, for human beings here on Planet Earth, the universe is, to say the least, a complicated place. Looking up — outward, really — we see the sun, the moon, planets of all kinds, then stars, myriad stars; they are born, they die; all kinds of stars, even galaxies of stars. We see our own galaxy, but from the inside. Beyond, there are more galaxies, in fact clusters of galaxies, clusters of clusters, even larger arrangements, structure upon structure, a "hierarchy of structure" to quote Vilenkin. And then there are black holes and, putatively, "dark matter", even dark energy! Plus some really exotic stuff. So how did this all start? Or did it? And when will it all end? Or will it? Even though our astronomers and cosmologists have made enormous progress, we don't really know. We don't know the *geometry* of the universe; not really, not for sure. We don't even know the *kind* of geometry for sure. Some say we don't even know if *our* universe is all there is. And let's not get into the meaning of "is"!

Back here on Earth, it's a bit different, though still very, very complicated, just in a different way; looking inward, we see cells, countless numbers of cells and great varieties of them; we can see their constituent molecules; we can even "see" the atoms of which the molecules are constituted although here it gets rather murky. Nevertheless, we "know" (What we believe to be authoritative physical experimentation tells us) that at an even smaller level there are things we call particles, some complex, some elementary, some fleeting or ephemeral, some stable and long-lasting to one degree or another. And, to close the loop; ultimately, aside from the dark matter and dark energy lately imposed upon our cosmologists (a big "aside"!) that's what the universe is made of. But, in contrast to the big picture of the universe, the particle picture seems to be very orderly. There are only discrete varieties within each of which all the members are identical. So why is that; how did *it* get started?

In a way, what I've been describing here is a parallel to Gulliver; the (immensely difficult to contemplate) universe we inhabit on the one hand and the (realistically impossible to picture) microworld of which we are composed on the other. So what I'm trying to say is that the best we can do with our theories, our models, our grand summarizations or our detailed itemizations of either the large or the small is to use what amounts to *metaphor* to propound *allegorical* patterns of what we perceive, structures we can understand, that appear to be self-consistent (according to how we define such criteria) and that we can use to make predictions (or post-dictions). We can do this in words, or in pictures, but inevitably in mathematical relationships that some say are all we can ever hope to put together, in fact that constitute the ultimate reality. Well, I think, maybe not quite; I would say, instead, that the *mathematics constitutes an indispensible metaphorical tool, a language that, in translation, also provides* an *allegorical, albeit a more formal, portrayal.*

Now there exists, as I'm sure you're well aware, what's referred to as the Standard Model of the elementary particles, a tightly organized scheme which has achieved the iconic status of received wisdom. In this book, as you'll see, I have managed to find myself in the precarious position of being skeptical of certain aspects, unfortunately rather fundamental aspects, of the SM for which I have developed some alternative notions. Broadly speaking, that's what the book is about but I hope you'll find it less narrowly focused than that. There's also what's sometimes referred to as the Standard Cosmological Model but, despite intense effort, it has only achieved the status of a working model, a situation that is completely understandable given the immensity of the subject. And our inability to subject it to experimentation. In any event, I must reluctantly describe each of these efforts as also fundamentally allegorical and by that, I don't mean to disparage them in any way because they are monumental achievements; it's just the nature of the game we are restricted to playing.

So, what's *my* "allegorical" point of view in this book about "*Knots, Braids* and *Möbius Strips*"? In this book, I maintain that the best allegorical picture we can devise all comes down, inevitably, to *geometry* which, I claim, we do know enough to talk about in some detail. In essence, that's what this book is all about — the geometry of the elementary particles. However, it's not as though that geometry is just another particle attribute. Nor is the book a minor variant or extension of that currently iconic, paradigmatic Standard Model of particle physics, definitely not that. In this book, particle geometry (within which we must include its topology) is the *whole thing*, the wellspring from whence emerges *all* the attributes, all the descriptors that, for us, define a particle — its mass, its electromagnetics, its spin, isospin, its various invariances, even its interactions with other particles, all just *manifestations* of its geometry.

And one more thing: geometry implies *structure* — configuration might be a better word — and the structure/configuration associated with the geometry I'll be talking about is highly visualizable. I think that's a good thing. However, we must not forget that the pictures and diagrams we'll show are, again, only allegorical. In keeping with my thesis, expressed above, they are only *metaphors* for reality and that's the best we can do; I don't pretend to know what the entities they portray "really" look or act like. On the other hand, like Feynman diagrams, they seem to function just fine to illustrate the ideas of the book and to guide the analysis thereto.

The eminent Physicist, Professor John Archibald **Wheeler**,[1] rightly celebrated for his numerous contributions to many avenues of physical research, was also known for his knack of encapsulating complex subjects in a few terse, pithy words or phrases. Of General Relativity he is reputed to have asserted something like "Matter tells space how to curve and spatial curvature tells matter how to move." Terse and pithy, indeed, but from the point of view of this book, not *quite* right. As written, there's an unspoken assumption in the professor's summary of GR: it is that matter and space are

[1]Whom you'll meet again later in the text.

distinct; that they're different kinds of entities; and that, in their interaction, matter somehow exists and moves against the *backdrop* of space. Or, as sometimes characterized, space is the stage upon which material objects act and interact.

Contrastingly, in this book, the elementary particles (of which matter is, of course composed) are viewed as localized distortions, *in-and-of* Space and Time, or better (since **Einstein** and **Minkowski**) in-and-of *Spacetime*. In fact, as we shall see, the elementary particles exist as *Solitons* [2], *continuously, seamlessly* forming and reforming of the stuff of spacetime even as they move along within it.

Actually, this is not a new or unique point of view; from a historical perspective, an almost exactly similar description harks back a century and a half or so (well before the relativistic unification of space and time) to the mathematician William Kingdom **Clifford** [3] otherwise famous for the algebra employed by Dirac to formulate his theory of the electron. Indeed, although he did not use that precise language, Clifford even viewed the motion of matter as the progression of a *solitonic* distortion through space. Had he extended his point of view, he might have made the connection, because the motion of a *solitary disturbance* moving through a fluid is well known to have been reported upon back in 1834 by the engineer **Scott-Russell** who happened to observe such a disturbance being initiated in water by the sudden cessation of the forward motion of a canal boat. Scott-Russell's observations are nowadays cited as the earliest mention of a solitonic disturbance.

The distortions I will be talking about take the form of Möbius strips, themselves definable as concatenations of the most elementary of a variety of *knots* known as ***torus knots*** (knots that can be *thought of* as strings wound around a torus). The solitonic nature and topology of such entities will be explored herein in terms of ***differential geometry*** in one form or another. And how they give rise to the complexities of elementary particle physics in terms of ***algebraic geometry***. In summary, that's the kind of localized but dynamic geometry the book is about: a combination of topology, algebraic and differential geometry, as required to demonstrate how all the attributes that define an elementary particle, the taxonomy stemming from the combinations of particles, and the interactions they engage in, are geometric in origin.

That may sound like a radical point of view but there is ample historical background for it too. For one thing, the notion that the fundamental constituents of matter might be extensive anomalies (rather than point-like singularities) in an otherwise featureless environment is not without precedent. In 1867, W. H. Thomson, Lord **Kelvin**, proposed [4] that atoms could be described as knotted vortex tubes in the "ether" then regarded as the medium for electromagnetic wave propagation. The idea attracted considerable interest for quite some time but was fated for eventual abandonment; it could not have been known at the time that atoms are composite.

Well, don't give up, I'm coming close to ending this preface but before I do, I want to make sure you know what I consider to be the crux of what we'll be talking about; it's not just Geometry, per se: it's the *right* geometry. And the reason I'm writing this book, the reason I pressed ahead for quite a long time developing the notions that are going into the book, is that I'm confident I've got the geometry right. I'll talk about how important that is a bit more in the Introduction, but I can indicate right here, why that's the case here:

In modern times, much effort, in fact, I maintain a great deal of *unnecessary* effort, has gone into mitigating or avoiding basic problems ultimately traceable to the inevitably *singular nature of point-like elementary particles.* The resulting theory is extraordinarily complex and burdened with a certain lack of grace. Today string theory, in one manifestation or another is generally regarded as a likely successor to the currently paradigmatic Standard Model of particle physics, in good part because the objects of interest (strings or higher dimensional manifolds) are nonsingular [5]. Earlier, in 1953, **Enz** [6], in analogy with the motion of Bloch wall discontinuities in magnetic materials, associated mass and size to a relativistic "particle" extensive in one spatial dimension plus time and featuring an internal angle variable, thus evoking a comparison with the *twist* of a Möbius strip. Enz also took note of the singularity problems associated with point-like particles and suggested that "...the idea of a particle extending in space should be favored, i.e. a particle with a *structure*."

That's my emphasis here because I agree wholeheartedly, as you'll see as you read on! And the fact that the particles we'll be talking about are solitons is indeed important but that's not the whole story either. Solitonic disturbances now known as Skyrmions were introduced by **Skyrme** in 1962 [7] and have found considerable utilization in, for instance, devices that employ magnetic materials wherein the orientation of magnetism varies in a cylindrically symmetrical fashion. Their connection to elementary particle physics is primarily through their identification as baryons but the geometry involved in any case turns out to be quite different from what will be described herein. More pertinent for present purposes are a pair of computer simulations one by **Ricca and Berger** in 1996 [8] which deals with stable, knotted structures that appear in fluid flow and the other by **Fadeev and Niemi** in 1997 [9] that exhibit stable, solitonic knot-like solutions to the field equations describing closed-loop strings. In each case the two simplest *torus knots* (to be discussed in what follows) were displayed.

There is much more that could and possibly should be said to set the stage for the subject matter of this book but perhaps I've already talked too long and it would be best just to begin.

I

Introduction: Some History and Philosophy

"Physical concepts are free creations of the human mind and are not, however it may seem, uniquely determined by the external world"

(Albert Einstein, "Evolution of Physics", 1938)

Indeed! And by physical concepts what the illustrious Professor Einstein meant, one might surmise, are certain elements, the body of which constitutes (in accordance with the Preface of this book) an allegorical model of that external world — formally, its Physics.

(Jack Avrin, "Knots, Braids and Möbius Strips, The geometry of elementarity", 2014)

1

History and Philosophy

In a way, this introduction is sort of a continuation of the Preface so you will be faced with a lot of words. However, where the preface was mostly about point of view, what we have here is mostly about context, like it says, historical and philosophical and by philosophical I mean some epistemology and some ontology — a little dose of E and O every once in a while does no harm I should think.

I realize that not everyone who reads this book is likely to take everything that's said in it at face value; however I trust there's enough of substance in it to at least set the skeptics to thinking — I would settle for that. At the same time, most of us would agree, I suspect, that the ancient maxim "first things first" is good philosophy — even for reading a book. So, let's start with the title; it's a bit cryptic but I hereby stipulate that the book's principle subject matter does, indeed, consist of knots, braids, Möbius strips and the elementary particles of physics, all inextricably linked into one harmonious whole, hopefully to the reader's satisfaction. That knots and Möbius strips appear together is not too surprising since both are longtime elementary objects of topological study (but there's more to it than that, as I shall demonstrate). Their relationship to elementary particles is, however, another matter, so crucial that without it, there's really no point to writing the book! Why and how that relationship comes about is a rather convoluted tale I also intend to unravel in considerable detail as the book unfolds.

Most of that detail derives mainly from material published over the period 2005 to 2012 [10–14], although I really started thinking seriously about the whole subject back in the fall of 1996.

Nevertheless, it's only gradually that I began to feel I could discuss it in a more comprehensive way and it took a while before I finally realized that there's a rather elementary topic that needs to be clarified, to wit, "just exactly what *is* an elementary particle?" Or, for that matter, what we mean by "elementary"? Or, even "particle". To say nothing about what we mean by "is" (As per the message of the Preface, not really a spurious matter!). The celebrated nobelist, **Eugene Wigner**, who did so much to advance the role of symmetry in physics is credited with a detailed *algebraic* analysis of the subject with the conclusion [15] that an "elementary particle 'is' an irreducible unitary representation of the group, G, of physics, that is, the double (universal) cover of the Poincaré group of those transformations of special relativity which can be continuously deformed to the identity."

Most impressive, but not exactly what we would like to see here, in a book of this nature; although proven useful to some of the scholars among us, it provides scant guidance for describing, say, the nature of the particulate occupation of space, in some ontologically satisfying way — somehow, a rather unsettling state of affairs. But notice; Professor Wigner's definition illustrates, albeit in a rather extreme way, the point made in the Preface that an allegorical description is the best we can achieve. However, then he takes the mathematical "hard line" point of view — that the mathematics "is" the ultimate reality — to which I objected in the Preface. In other words, I do not think one should say that "an elementary particle ***is*** an irreducible, etc." but, to adhere strictly to my thesis, something like "can be viewed for analytical purposes as" would be more like it. However, I forgive the Professor since he (or somebody) put his "is" in quotation marks indicating that he was really thinking along those lines!

Well, it looked like some research was in order, so I looked in the dictionary! Here is what it says [16]: *Element*: "Of first principles: of the rudiments or fundamentals of something". *Elementary Particle*: "A particle smaller than an atom, which is capable of independent existence, as a neutron, proton, electron, etc." Those definitions seem to make sense, so, let us keep them in mind. But for *Particle*: "In Physics, a piece of matter so small as to be considered without magnitude though having inertia and the force of attraction", I am rather dubious about that so I am just going to recommend we forget it. Anyway, we'll get back to the subject of elementary particles soon but before we get too far, there is something we absolutely have to do: we must acknowledge the omnipresence of the now iconic, even paradigmatic Standard Model of the Elementary Particles.

The SM is clearly a magnificent edifice, constructed by the inspired contributions of many talented theorists and experimenters primarily over the second half of the 20th century. With the putative discovery of the long-sought "Higgs boson", its practitioners now consider the SM to be a complete theory and, in fact, many consider it *the* most important development of the latter half of 20th century physics, a characterization that might be disputed by others (for example solid state physicists!). I would not presume to venture an opinion thereto but, in previous publications I have had the temerity to label what was developing in my personal corner of the particle world as an "Alternative" to the Standard Model! On second thought, however (part of the comprehensive look alluded to above) that's not quite right. I think it's really more appropriate to describe it as something that "reflects the taxonomy, interactions and attributes of the SM but in a simpler and more efficient way"; rather more wordy, but more accurate.

Which might be judged as not much of a reduction in hubris! Nevertheless, I shall endeavor to justify that claim in detail as the book progresses. For now, it's only fair to ask what "simpler and more efficient" actually means. To begin with, there is no denying that the SM is a very complex theoretical structure with a pervasive basis in what's known as Quantum Field Theory. They write books about QFT, lots of them [e.g. 17], on a variety of levels and my level is probably best characterized as "neophyte". And, having gone that far, I might as well confess: I'm not really a particle physicist (which may become apparent as the book progresses anyway!). In fact, I'm not a physicist at all. Nor a mathematician. Although I had a couple of degrees in physics that was a very long time ago and what I'm now is a "superannuated, long-retired Defense/Aerospace Engineer".

When I was working, I was involved with the analysis and design of systems that gather, process and utilize information — radar, communications, guidance and control, that sort of thing. It was a very gratifying 40 years or so and I learned some valuable lessons applicable to life in general as well as to engineering. And, I maintain, even to fundamental physics. A most important lesson is that there is often more than one approach to solving a problem. Naturally, some approaches are "better" than others — more efficacious, we might say — often leading to less effort, less complexity, or even a better understanding of the issues involved; such situations abound at every level of activity. And, where basic physics, engineering or a multitude of applications thereof are involved, characterizing the basic geometrical and/or topological nature of the situation often turns out to be all-important. Given the right G/T, what's required usually flows naturally and simply whereas, if it's wrong or suboptimal, it often adds complexity and labor. Unfortunately, experience shows optimality to emerge usually only after arduous suboptimal pursuits — that's life. Even, as history shows, more often than not it's also the story of the physical sciences.

You know, nowadays the ancient astromomer Claudius **Ptolemaius** (Ptolemy) [18] is "dissed" (disrespected!) quite a bit. To say something is Ptolemaic is to criticize it as being encumbered with a lot of unnecessary detail that could have been avoided by taking another approach. As we know, the other cosmological approach was (finally!) supplied by Mikola Kopernik (Nicolas **Copernicus**) [19] who took his life in his hands to develop the radical (at that time) notion that the earth was not the center of the universe; the sun was! Now Micky was probably no smarter than Claude. In fact, from the point of view of numerical predictions, the Ptolemaic system was every bit as reliable, and sometimes better than the Copernican which was somewhat oversimplified. But Ptolemy's geometry ensued from what nowadays you might call a rather "eccentric" (!) point of view. And that's what made him work so hard adding more of those pesky "epicycles" to his theory every time a new batch of data came along in the tracking of the planets. The problem was that the basic ***geometrical*** *point of view* of his system was just ***wrong***. It had no future whereas, with some (actually rather important) tweeking, the Copernican Solarcentric geometry was ***right*** at least for the solar system (!) and was eventually adopted by all right-thinking astronomers. Astronomy (and Physics) was off and running with Kepler, Galileo, Newton *et al.* Of course, it was a little touchy at first because you could get burnt at the stake or forced to recant your ideas if you expressed them too loudly!

It is with that background in mind that this book is written. I might as well just come right out and say it (Stake burning being out of fashion): I do not claim to be a modern day Copernicus (I am not even Polish) but, in terms of the elementary particles of physics, I believe I've got the G/T right and, unfortunately, the SM does not. It seems to me that The Standard Modelers have been working much too hard to get where they got. And I'm not alone in harboring such a viewpoint; the practitioners of string theory evidently feel much the same way about their G/T vis-a-vis the Standard Model as attested to by some of the best known String theorists [5]. One reason I think I've got it right is that, as my research progressed, important attributes did in fact seem to emerge in a "natural and simple" way. Here's a partial list of such that I'll just mention without explanation at this point; it's very gratifying:

1. CPT invariance.
2. The ***manifestation*** of
   a. Nontrivial vector bundles
   b. The gauge groups, U(1) and SU(2) and their simultaneous validity.
   c. Spin (in multiples of 1/2) and Isospin.
   d. Fractional electric charge without Quarks.
3. A geometric interpretation of electromagnetism.
4. The relationship between charge and isospin (as per Gell-Mann and Nishijima).
5. The equality of electron and proton charge magnitudes.
6. The notion of antiparticles as particles moving "backward in time" (as per Wheeler and Feynman).

7. The taxonomy, interactions and attributes of the Standard Model.
8. Lorentz Invariance of basic particles.
9. A fundamental connection to General Relativity.
10. A relationship between particle mass and "size".
11. An estimate of the "size" of the electron..
12. Connections to

    a. Dirac theory.
    b. Topological Quantum field Theory (TQFT).
    c. Hopf algebra/Quantum groups.
    d. Quaternions.
    e. Instantons.

13. An "indigenous parallel" to the Higgs field of the Standard Model.

Some, maybe most, of the above will be seen to be related as we proceed. Also, that last point will of course be recognized as of considerable current interest and will be treated appropriately in that light; it turns out that there's some most interesting connections to other parts of physics lurking in that "indigenous parallel" business that you might not want to miss! Anyway, I trust I'm not getting ahead of the story here but most readers will probably have at least seen reference to quarks, gauge groups, etc. and I want to make a major point here:

In terms of being "natural and simple" the work to be described herein produces those results listed and more, but ***without recourse to the SM's quarks***; or to the ***gluons*** that sequester them within their nucleonic clustering; or the latest-to-be-added quantum parameter, "***color***" (and the theory, "***Chromodynamics***", and associated gauge group, **SU(3)**, that govern it) invoked to insure that the ***Pauli*** principle is not violated, and to explain how it is that such important entities as quarks have not, and in fact, cannot be seen!

Well, maybe we are getting a bit ahead of the story so let us go back to the meaning of "elementary particle". It turns out that for our purposes it's related to the epistemological paradigm of

Reductionism. As an example of that principle at work, consider something of interest to all of us: life on Earth. It is incredibly complex and varied. On a personal basis (to which I subscribe wholeheartedly) human beings are highly composite; the biologists tell us we are all made of "cells" but that there is an astronomical number and an immense variety of them. However, the biochemists tell us that cells are themselves constructed of only a score or so of molecular varieties. That's a lot of reduction right there. Furthermore, each molecule is in turn made up of only a very limited number of kinds of atoms, those in fact that constitute the elements essential to life (at least here on Earth but let's not get into that). Reduction squared!

So now we're down to the domain of physics, atomic physics, to be specific. Isaac **Newton** is reputed to have believed that all matter was ultimately composed of miniscule, very hard and indivisible "atoms". But it was the chemists, of whom John **Dalton** in the mid-19th century was the most prominent, who led the first attempts at systemizing the variety of elementary substances they encountered in terms of their atomic constituency. However, the term "Atom" itself is of ancient origin. Educated men in those days such as Newton and Dalton being schooled in classical literature were probably well aware of how the word, "***atomos***" originated with a pair of ancient Geeks, the philosophers **Leucippus and Democritus** in the 5th century BCE [20]. Not having had access to microscopes, Bunsen burners, etc., to say nothing of high energy particle colliders, L&D had to do a lot of heavy thinking. For instance, they said (I am paraphrasing, of course) that if you cut a block of wood into equal parts and repeat with each part and so forth, eventually, you will get to a smallest part that you will be *unable* to cut into smaller parts; it is "**atomos**". Which may be Greek to most of us but that's exactly what it meant — uncuttable. That's reductionism taken down to its limit — you might say, its "elementary particles"!

That's fascinating history but after Dalton (and of course, **Mendeleev**, and his Periodic Table) nothing much happened by way of *particulate* reduction (In fact not everyone believed in the existence of atoms) for most of the 19th century (with two major exceptions very relevant to this book; look for them in the next chapter!). Instead, especially with the work of **Faraday and Maxwell**, continuum physics, the subject of *fields* and the wave phenomena they implied was of paramount importance.

However, the advent of the 20th century signaled the beginning of a most heroic time for atomic physicists. First **Planck's** derivation of the correct blackbody spectrum and **Einstein's** treatment of the photoelectric effect showed that radiation acted as though it came in clumps — ***quanta***. Then **Rutherford's** definitive experiments showed that atoms were themselves composite; they appeared to consist of tiny nuclei "somehow" surrounded by electrons, a notion that **Bohr's** "old" Quantum Theory soon quantified; a major step toward reduction. And when **Chadwick** eventually showed that there were two "nucleons", that is the neutron as well as the previously identified proton, the composite nature of the atom was well on its way.

Then, of course, we had the quantification of angular momentum and magnetic moment, the invention of Quantum Mechanics by **Heisenberg, Schrödinger and Dirac**, the need to add the quantum parameter of "***spin***" (and the notion of bosons with integral spin and fermions with half integral spin), **Pauli's** exclusion principle and his realization that (what came to be called) the neutrino must be postulated to make the energetics come out right in weak reactions, and, finally Heisenberg's invention of even one more quantum parameter, ***isospin***, one that united protons and neutrons as being just two aspects of the same thing — if you neglected electromagnetic charge, that is. The reduction of the atom appeared to be complete and it was possible to develop a taxonomy of all the naturally occurring elements as a modernized version of the **Mendeleev's** Periodic Table, complete with a theoretical rationale.

So that's where matters stood at the advent of World War II and in fact, pretty much even afterward when I came back, finished schooling went to work in the missile guidance business in 1952 and started learning about things like radar and feedback control systems. By the way, you may find it hard to believe but both of those subjects, may have a relationship to the subject matter of this book! (See Sec. VI.) Where matters stood, that is, except for a few details (!) like nuclear fission and the development of greatly enhanced experimental techniques and devices. Nuclear fission of course led to intense study of the nucleus and the nucleons themselves. Nuclear physics and the modern era of high energy particle accelerators were upon us.

But there were also some devices capable of much refined measurement to come out of the war effort and experiments utilizing these capabilities soon came up with disturbing refinements of such parameters as the magnetic moment of the electron. Such micro-measurement anomalies were recognized as implying a major crisis in our understanding of fundamental physics. Its resolution led to the development of what became a major paradigm, **Quantum Electrodynamics** (QED), eventually to become the model for **Quantum Field Theory**. And, a most important but disturbingly complex theoretical *procedure* known as "***renormalization***". The subject of renormalization is also related to matters of geometry and topology but, while it is, indeed, extremely important to the Standard Model, it is not directly related to our reductionism story, so at this point we have to postpone talking about it.

What does relate directly to the reductionism story is another crisis that arose a bit later and it has to do with the ever-expanding capability of particle accelerators. At some point in the second half of the 20th century that capability began to yield indications of particles that had never been seen before, with characteristic combinations of creation and decay that were referred to as "strange". The problem was how to include them in the reasonably well-understood taxonomy of that day or, more likely, how to create a more comprehensive way of systemizing. What happened then is a familiar story: two theoreticians, **Murray Gell-Mann and George Zweig**, working independently, took the very successful reductionist route and postulated the existence of three varieties of (what became known as) "***Quark***" to make up all known hadrons whether fermions or bosons and including the strange new particles. (Actually, Zweig preferred the name "Aces" but lost out to G-M's persistent insistence!)

A crucially important, revolutionary new feature in their theory, it turned out, was that the quarks had to be endowed with an unprecedented ***fractional electric charge***! As you no doubt recall, it is two "up" quarks (each with charge number +2/3) and a "down" quark (−1/3) for the proton and two downs and an up for the neutron with all charge numbers as multiples of the charge of the proton — you do the math — and a quark and an antiquark for mesons. As for the "strange" particles, an additional "strange" quark was added to the mix to replace one or more "down" quarks in their makeup. Eventually, as you will no doubt also recall, three more quarks were added to the mix to make up three total generations in the family tree; that is three doublets, Up–Down, Charm–Strange and Top–Bottom.

This was all in or around 1964 but meanwhile, back in 1948 the illustrious **Enrico Fermi** and the soon to become illustrious **C. N. Yang** wrote a short, almost apologetic paper [21] suggesting that the pions (which are bosons) might be bound states of a nucleon and an antinucleon. Although, that paper didn't get an awful amount of play right at that time (but look for such combinations later on in this book), about the same time as Gell-Mann and Zweig's introduction of quarks, another pioneer, **Soichi Sakata** in Japan, was also thinking about what to do with the influx of new particles but in a completely different way. According to [22], what Sakata did was to extend Fermi and Yang's ideas to create his own systemization scheme [23]. His innovation was to consider all the particle's that were otherwise built upon quarks by G-M/Z, but to portray them instead as **bound states** of just **three well-known basic particles**, all fermions, in this case the two nucleons and the so-called Lambda particle (which was itself "strange").

In other words, instead of embodying another phase of *reduction*, Sakata's scheme was ***self-contained***; that is all particles, basic as well as those derived by ***combinations*** of the basics were included in the resulting taxonomy and interactions. This scheme, which I shall henceforth refer to as the Fermi–Yang/Sakata (FYS) model was thus ***self-sufficient***; there was *no* ***need*** to invoke another level of reduction! Unfortunately, although that model looked pretty good for a while, at some point it ran into some difficulties that could not be explained soon enough (and I cannot explain them quickly here, either). To make a long story short, as we know, the quark theory won the day; reductionism was once more triumphant. So, where will it all end? Or will it?

Well, at long last, that brings us back to the story of Knots, Möbius strips and Elementary Particles and, in particular, to the Gell-Mann/Zweig vs. Sakata story and to my conclusions thereto, which are as follows:

- I believe **Sakata** was on the right track, the FYS track.
- But **Gell-Mann** and **Zweig** were *almost* on the right track too!
- What I mean is that Sakata was right about the notion of ***self-containment***.
- And G-M/Z were right about the need for ***fractional electric charge***.

The reason I believe the validity of *both* of those apparently incompatible approaches is, briefly:

- because the ***geometry/topology*** of my elementary particles ***inherently manifests*** fractional electric charge.
- And at the same time, the ***taxonomy*** that results from that kind of elementary G/T also ***manifests*** the epistemological approach of the FYS model (although not with quite the same basis).

Philosophically speaking, from an ***epistemological*** point of view, the elementary particle model to be discussed herein (Despite what I said above, I shall continue to refer to it as the "Alternative" Model!) thus represents an ***amalgamation*** of two very diverse approaches (GMZ and FYS) to turning a major crisis in fundamental physics into a major advance. To me, that's a most gratifying accomplishment, similar in spirit if not in substance to some of the other combinatorial advances we've witnessed in the history of physics.

And while we are in a philosophical mode, what is definitive about the "Alternative" ***ontology*** in this book is that it is based on ***Toroidal Topology***; the topology of the torus. All the good things that emerge — the attributes and features I listed some pages back — in a fundamental sense, owe their existence to the fact that the associated "particles" (that we shall define) are built upon what might be described as a toroidal "***last***" (in case you are not familiar with that term it is the form upon which a shoemaker constructs a shoe — at least my understanding is that that is the way they used to do it!). We will talk some more later on about some of the things that make torii so special but, right up front, we can cite a fundamental issue, namely that the ***group structure*** that governs our particles, and their ensuing taxonomy derives directly from their toroidal topology.

Before we finish up with this Introduction, I want to show you a couple of figures: Figure 1-1 encapsulates in a very broad sense what the thesis of the book is about. Also, Figure 1-2 is intended to show how the subject matter of the book interrelates with other topics of particle physics (GR means General Relativity and TOE means Theory of Everything!). I will try to make those connections as the manuscript progresses but you can check back with the figure to see how I did!

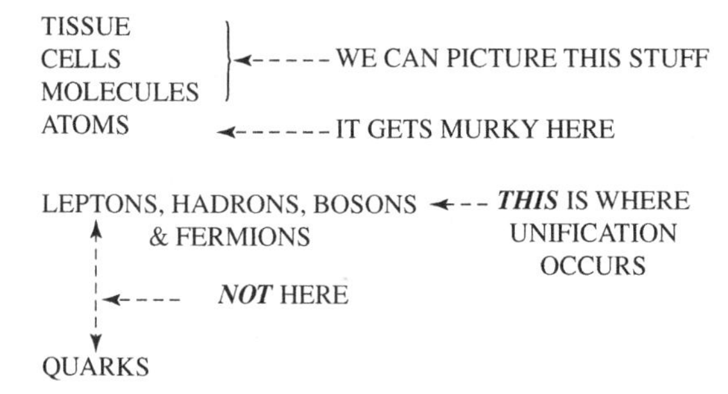

Figure 1-1. A very broad view of what this book is about.

Well we are almost done but of course, as always, there are some details that need to be mentioned. Which brings us to the next section, the one that actually begins the discussion of how knots and

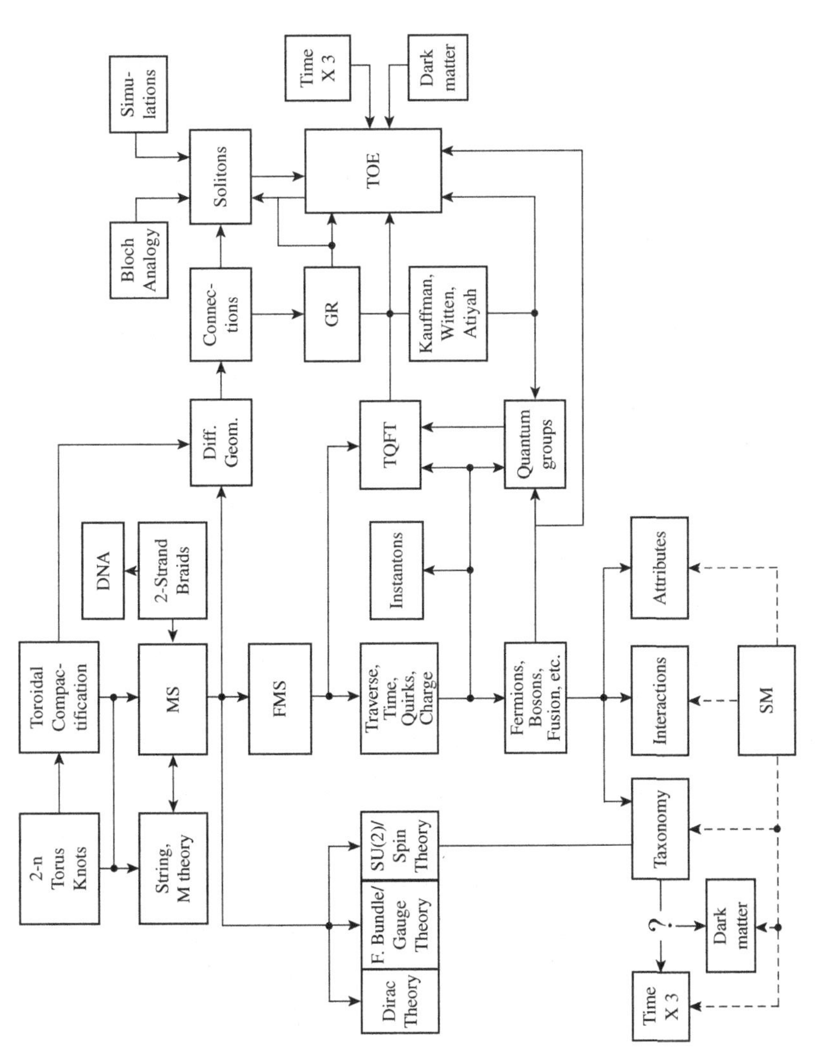

Figure 1-2. Connections (some tenuous) between various topics in fundamental physics.

Möbius strips are related and how both are related to elementary particles. But, looking ahead a bit, as you might expect from the preceding, these are not the kind of particles your grandmother was used to. Nor are they quanta, each of its own field as in the Standard Model. In this book, ***there is only one field***, **Spacetime** itself. I repeat: ***only one field***, **Spacetime** itself, of which our elementary particles are formed and within which they move and interact. Our particles are to be regarded not as objects *distinct from the backdrop* of that spacetime but as distortions *in-and-of* it. They are ***Solitons***, *self-perpetuating* disturbances, of a special kind, with their own unique Geometry/topology [2].

As mentioned in the Preface, this point of view was essentially anticipated by **Clifford** [3] a century and a half or so ago, decades before General relativity, although he was aware of and greatly impressed by **Riemann's** work on non-Euclidean spaces. We shall also be concerned with Riemannian disturbances but on a local basis as described in the Differential Geometric part of the book which, according to the organization of the book, will take us a while to get to (But don't give up!). The plan of the book is a division into ***sections*** which are further subdivided into ***chapters***.

The next section (Sec. II) has to do with introductory and basic concepts and what I call the "Algebraic Geometry" of our version of an FYS model meaning things like Taxonomy and the Interactions of particles, basic as well as composite. At that point the treatment is generic in a way not yet explicitly identified with the Standard Model but Sec. IV does make that all-important comparison. (The reason for the delay is explained therein.)

Section III is interjected as something of a hiatus wherein I take some time to look at several algebraic subjects that are not *exactly* part of the main story line but important nevertheless. The first chapter therein shows that the Alternative Model's taxonomy constitutes a Hopf Algebra and the second chapter is a foray into Quantum Mechanics in a rather unusual way. We then go on to talk about spin and several topics I trust you'll find interesting and informative (including Quaternions and Sample data feedback systems!).

As stated above, Sec. IV does make the comparison between the taxonomical development in Secs. I and II and the Standard Model. It includes a final chapter called Family Matters that addresses the existence of those still somewhat mysterious additional "generations" of particles encountered only in high-energy particle accelerators beginning with "Strangeness" as in the above. As we know, the SM copes with the existence of additional generations simply by adding four successively more massive quarks for a total of six. In this book, however, building on the way our particular FYS model endows the basic particles with electric charge (Described in Sec. II as a U(1) rotation between space and time), an admittedly speculative geometrical argument is developed to justify expanding our view of spacetime from a 3+1 manifold to a 3+3 manifold, three dimensions of time as well as three of space. In other words, to justify the *triplication* of time, thus putting it on essentially the same footing as space and enabling the concept of reorientation within the time submanifold as well as in that of space.

With Sec. V, we get more physical (There really is some Physics) beginning with the Differential Geometry of the basic particles. We then go on to show how the solitonic nature of the basic particles of the model arises and to verify an earlier definition of antiparticle, to generate an estimate of the "size" of the electron and, finally, to develop what is described as "an indigenous parallel" to the Higgs field of the SM but in a way that is associated with the spacetime structure of each basic particle rather than, as is the case with the SM, with the potential energy each particle experiences as a resident in space. The section ends with what I consider a most intriguing connection between General Relativity and Quantum Mechanics; I will be interested to see how that plays out on the big screen.

Section VI is devoted to a variety of topics not particularly restricted to any of the other sections. It turns out that there are a number of intriguing connections to other aspects and theories of fundamental physics to discuss (recall Figure 1-2, above) as well as some possibilities for more investigation. Finally, Sec. VII is intended, first, to recapitulate what was discussed earlier in a way that puts the various topics in some kind of relationship to the rest of fundamental physics and then to present (my understanding of) current views of Cosmology in a rather personal way that includes a highly-condensed version of work by another author that may, at first glance, not have much to do with Cosmology at all (but you be the judge), and finally, the extension of some formalism developed in Sec. III to a speculative connection between knots and the cosmos.

The book ends with some concluding remarks and a number of appendices as listed in the Table of Contents; generally speaking they extend discussions treated in the main body of the book. However, I urge you to not overlook Appendix F. That Appendix, the last in the book was appended, almost as an afterthought, in deference to what I deemed the need for timely publication of the book. Its genesis stemmed from my realization late in the writing, that I had neglected a possibly novel connection between two (seemingly) totally disparate aspects of the physical world. The result was the emergence of a fundamental principle underlying the structure of that same world, one uniting the bases for such diverse fields as Radar, Quantum Mechanics, Biology and Cosmology. It is all about Complementarity and, in keeping with the title of this chapter, it even appears to have a fundamental connection to the ancient philosophy of the Orient! Who knew? At any rate, it is also discussed at some length in Chap. 43 of Sec. VII.

We now leave the realm of History and Philosophy and begin the description of a system of elementary particles, one that, as hinted at by the subtitle of the book, constitutes an Alternative View. Enjoy!

II

Basic Concepts: Algebraic Geometry

"The whole of Science is nothing more than a refinement of everyday thinking"

(Albert Einstein, Physics and reality, 1936)

True, but easy for him to say! Such refinement often takes a lot of effort by a lot of people, (not all of whom are similarly gifted) and over a lot of time.

(Jack Avrin, 2014)

As per the Preface and Introduction, this book is about how the elementary particles constitute localized distortions of Spacetime and we are concerned with the geometric nature of that distortion. Broadly speaking, we can discuss the subject in terms of both the ***algebraic*** (essentially discrete) and the ***differential*** (continuous) aspects of that geometry. While this section emphasizes the ***algebraic*** aspects, as we will see, it is probably just as much about how geometry can be used to elucidate the algebra and vice versa.

2

Genesis of a Particle Model

In this chapter, a small set of particles is defined to serve as the basis for what constitutes a particle model, the Alternative Model (AM) that, eventually, we can compare to the Standard Model of particle physics. In the spirit of the FYS model, these particles, in combination, are relied upon to generate all the rest of the model's taxonomy, interactions and attributes all of which will be seen to be highly *visualizable*, a factor that lent encouragement to the pursuit of the model's development. In fact, before we describe the model in any detail, it may be revealing to recount an incident, you might say an ***epiphany*** of sorts (but without spiritual connotations!) that, for me, initiated the process of model development. In the waning years of the 20th century I was staring at the cover of a small book on recreational topology [24] published in 1964 (by coincidence, you will recall, the same year Gell-Mann and Zweig brought out their quark theory).

Figure 2-1 shows what the cover looks like: no printing, no information at all other than the pictured diagram, which clearly depicts a ***flattened*** Möbius strip (**FMS**), in fact, the canonical, flattened, one-half-twist MS, with a minimal, triangular planform ubiquitous in elementary topological disquisition (and commercial logos; you've undoubtedly seen it).

But now, suppose we augment that FMS just a bit, arbitrarily associating a direction of ***traverse*** to it as in Figure 2-2.

Then we can characterize the corner folds as *two* ***down***, *into the plane* of the diagram and *one* ***up*** *out of it* (think of the fingers of your right hand). And if we start out with an FMS with the

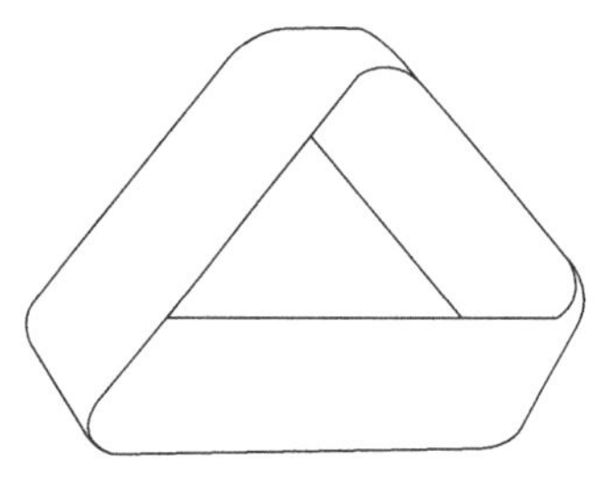

Figure 2-1. The FMS on the cover of that 1964 book.

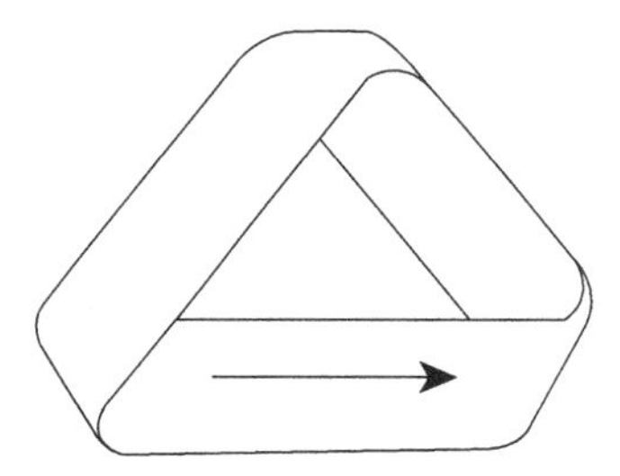

Figure 2-2. Same FMS with traverse.

same direction of traverse but with the *opposite* direction of *twist*, we end up with *one down and two up* folds. Does that remind you of anything? I thought it might! And, that being the case from now on I'm going to call the folds "**Quirks**" (not Quarks but **Quirks**!). Not only that, but if we start out with *three* half twists, we get three ups or three downs depending on which direction we twist — verry interesting, no?

It is readily seen that the triangular planform is the minimal ***flat*** configuration that can accommodate the twists of a Möbius strip and that there can be only four such, two with one half-twist (both left and right twist) and two with three half-twists. I call these the ***basic*** *set* and here they are in Figure 2-3 with left-handed twist on the left and right-handed twist on the right. The number of ***half twists*** (NHT) for the upper two is NHT = 1 and for the lower two it is NHT = 3. Letters u and d signify the up and down quirks, respectively. From now on the labels A, B, C and D will represent these diagrams until we make the connections to the Standard Model.

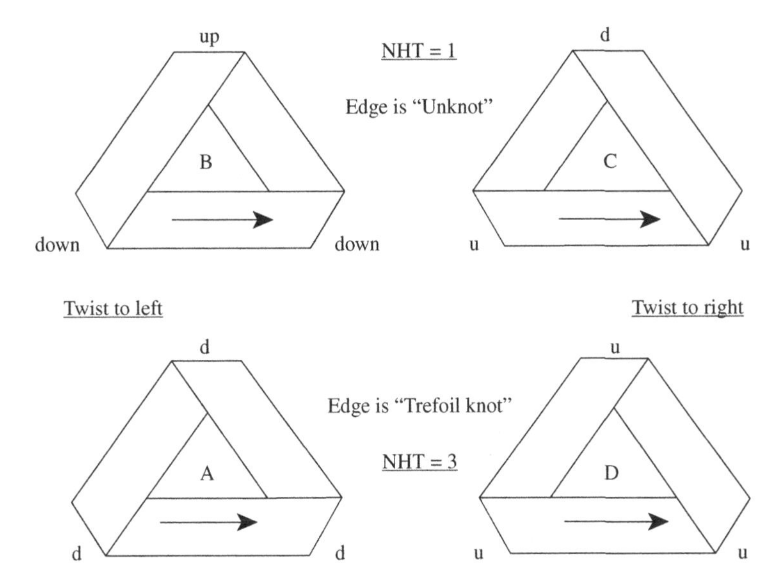

Figure 2-3. The basic set of FMS.

(Actually, for completeness, I also ought to include the closed strip with no twist at all! (But I will talk about that later on).)

If you were to make an MS out of a strip of paper (better, I would recommend a soft leather belt), you would perform a 180 degree twist to the right about an axis running down the middle of the strip before joining the two ends in order to have an NHT of +1. In what follows, ***twist*** (as denoted by NHT) is a most important parameter, which, as we shall see, is treated herein as a *manifestation* of the physical parameter ***isospin***. Also, note the left–right *mirror* ***symmetry*** of the figure, broken by the choice of traverse; it's something to keep in mind as we proceed to illustrate the development of a taxonomy. And, in that regard, as we shall see, arbitrary values of MS twist can also be realized as *composites* of these four basic triangular FMSs. Something to look forward to as well.

But how about ***knots***; where do they come into the picture? Knot Theory is of course a massive subject in its own right as a part of mathematical topology. A glance at the titles listed as part of the *Series on Knots and Everything* as well as the subjects included in the *Journal of Knot Theory and Its Ramifications* will confirm the scope of its involvement in scientific pursuits. However, in this book we are concerned with only a miniscule part of that scope. If you've absorbed most of what was discussed above, you are probably prepared to believe the following: as is well known and readily verified, the ***boundaries*** of the four basic FMSs can be viewed as ***knots***, specifically, the folded-over unknot (just a circle) and the trefoil knot for the single and triple half-twist MS, respectively, the two most elementary members of the genus of ***torus knots***.

For example, in Figure 2-4, the figure on the left is a *partially-flattened* representation of an MS for which the number of half twists, NHT = 1 (as in C in Figure 2-3) and on the right we see the "doubled-over ***unknot***" with one crossover that forms the MS *boundary.* Clearly, *two* traversals in real (three) space of the polar orientation vector in the figure are required in order to return to an original location on the unknot loop in this doubled over condition. We shall have a little more to say about that requirement soon in a little more formal way (there's a connection to group theory).

Actually, the boundary is representative of all similar paths on the MS. Thus, another way to look at it is to view the MS as a strip or ribbon consisting of a parallel set of much narrower such strips. Upon twisting and recombining, the leftmost (say) and rightmost subsidiary strips connect to form a torus knot as do the next such pair, and so forth, to form the complete MS in what amounts to a

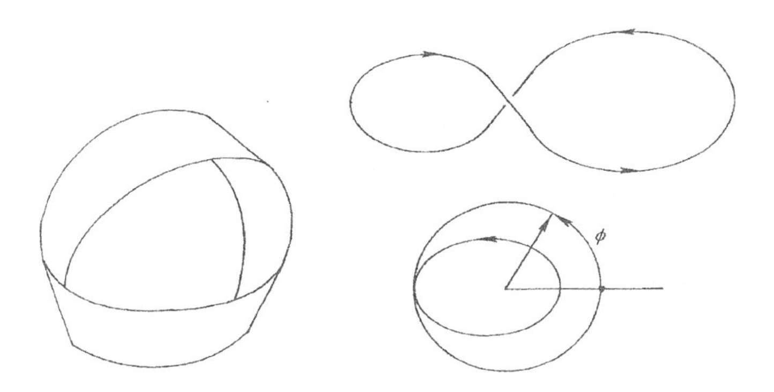

Figure 2-4. Single half twist MS (NHT = 1) with boundary.

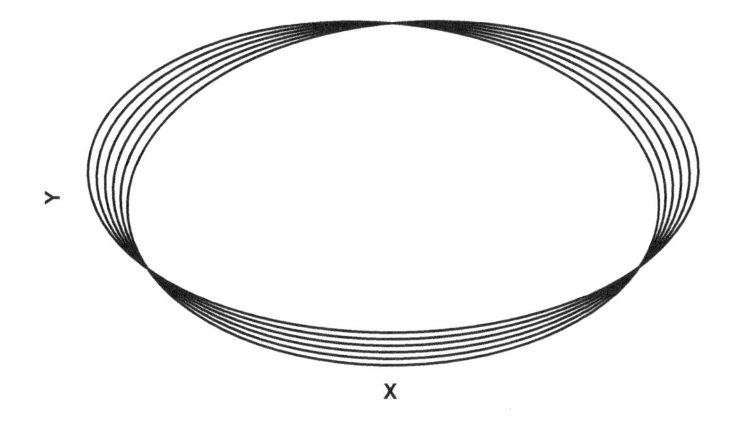

Figure 2-5. An NHT = 3 MS as a concatenation of trefoil knots.

concatenation of such subsidiary torus knots. Thus it also takes ***two traversals*** (but see below) of *that* **MS** to bring an asymmetric, oriented figure back to its starting location with the same orientation.

Consider, for example, the MS with *three* half twists (NHT = 3): in this case the knot in question is the trefoil knot, which is knotted so as to have three crossovers. Figure 2-5 shows the *discrete concatenation* of three trefoil knots (an arbitrary choice; it could have been four or five, or whatever, rather than three such knots) to form such an MS (it is the same figure you saw on the cover). What look like three nodal points are really just where the plane of the strip is oriented *normal* to the plane of the picture (we shall have more to say about that later on in this chapter). Also note that the MS looks like it's 6 knots wide (in general the MS would look like it's composed of twice the actual number of constituent knots; that is because of the two traversal requirement). Finally, concatenation is *one* way to accomplish what is termed the "framing" of a knot. Actually, as we shall see, all Möbius strips, flattened or not and basic or not, can be viewed as "framed" (2, n) torus knots (see below for definition). Thus, the essence of our particle model is seen to reside in such knots and in their associated ***toroidal topology***. As we shall see, the notion of torus knots and their concatenation is useful in deriving some numerical results of analysis (see Sec. V).

Interjection:

1. *A Bit of Propaganda:* It is this toroidal topology that is indeed the topology mentioned in the introduction as being fundamental to minimizing the complexity of developing a model of the elementary particles! By way of contrast, for hundreds of years, (you might say 2.5 millenia!) as noted therein, the ***default*** elementary topology has been that of a point or, at least a very small, spherically symmetrical region within which there resides (*somehow*!) all the attributes required to make the theory work. That topology has been the source of much grief as well as theoretical complexity needed to surmount the difficulties induced therein (but correspondingly a number of Nobel prizes!). Perhaps, it's time to take a different approach. In fact, string theory constitutes one such and this book features another that has some relationship to that intensively pursued theory. However, there are some very important, fundamental differences such that the two subjects are, effectively, completely diverse. Anyway, we will talk about that later too.
2. *A Touch of "Arcanity":* As noted above, Möbius strips and the associated torus knots, usually appear in the role of demonstration devices. In more arcane language they are referred to as ***manifestations*** of the 2-to-1 ***cover*** of the group SO(3) of rotations in three-space by the gauge group SU(2). That gauge group is representative of ***all spinors*** designated by ***odd*** values of integer, n, as it appears in the quantum value $n/2$ ***as well as*** all vectors designated by ***even*** values of n. Thus, we emphasize, once more, that the MS genus as well as the genus of $(2, n)$ torus knots of which it is constituted ***manifest*** the gauge group SU(2). It is interesting that, in harking back to **Wigner** and his definition of an elementary particle, mathematician **Shlomo Sternberg** in a book devoted to the connection between group theory and fundamental physics [15] infers, in passing, the suggestion that ***all matter*** "is built up out of spin-1/2 particles"! As we shall see, this book purports to constitute a ***realization*** of that suggestion. We might also mention that nowadays gauge groups

and their differential geometric characterization are most often carried out in terms of ***Fiber Bundle*** theory [25] wherein you often see reference to notions like ***holonomy***. In elementary terms, manifold holonomy can be characterized by how much a vector fails, upon transport around a closed curve on the subject manifold, to return to its original orientation while maintaining it's original attitude. A ubiquitous example is the surface of a sphere and the curve is a spherical triangle all of whose angles are right angles. In the case of a Möbius strip, the holonomy is basically just its ***twist***!

It looks like we are getting ahead of ourselves again so we had better go back and talk about torus knots a bit more. Here's what we mean by a torus knot: In our macroscopic world it's basically just a string (that can be *thought about* as) wound around a torus before its ends are joined. However, a ***caveat***: Our microworld torus here is implicit and ephemeral — it's not really "real". If it were, there would be no way to free it from its knotted imprisonment without drastic surgery! Of course our "string" is not really "real" either in the sense of the tangible thing we're used to in the macroscopic world. However, it can be *thought of* as a one-dimensional entity characterized, as we shall see, by a linear energy density — a "stress". Nevertheless, it helps to *assume* an explicit toroidal geometry in order to perform some analysis (which we do later on in Sec. V on differential geometry) so Figure 2-6 shows a segment of that kind of geometry.

Restricting the knots to be of genus $(2, n)$ then means that, with reference to the figure, the knot must complete n circuits in the meridianal, or θ direction for every 2 traversals in longitude, ϕ. Of course, the relationship is not immediately obvious between, on the one hand, the geometry of torus knots as strings wound around a torus and, as per the above, the MS associated with them, and, on the other, the geometry associated with ***flattened*** MS but we shall get to that matter shortly. In the meantime, we note again that it is only for ***odd*** values of n that we get a knot; *even* values create links (which, in this book, translates into ***odd values for***

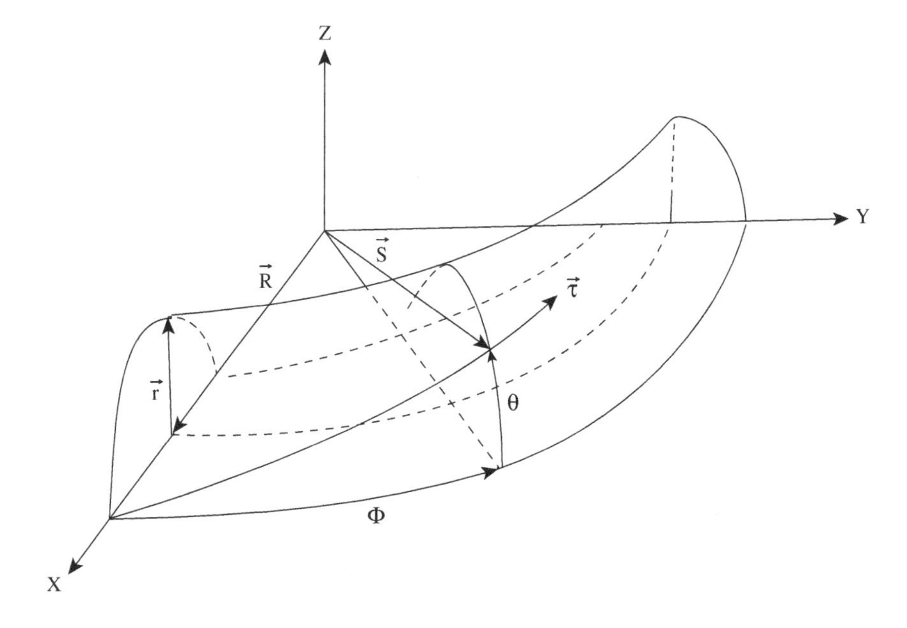

Figure 2-6. A segment of a torus knot.

Fermions and ***even values for Bosons***, respectively). A graphic way to show this is to depict the equivalent situation involving what can be described as two-strand ***braids*** with ***closure***, meaning that the top and bottom of the strands, indexed left to right as depicted in the first column of Figure 2-7, are connected (the negative sign just means twist to the left). We note that **Alexander's** theorem [26] says that ***every*** knot can be viewed as a braid with closure but, although there are of course an immense number of braids, we will be concerned here only with two-strand braids and only a limited number of those.

As we see in the next column, the result of closure for $n = -1$ is the folded over unknot with one crossover while $n = -3$ gives the trefoil with three crossovers. On the other hand, for $n = 0$ we get a pair of unlinked loops while for $n = -2$, the two loops are linked together at two crossovers. These are just examples of the general rule: *odd n gives knots. specifically torus knots while even n gives links.* Again, the number of crossings clearly correlates with knot/link parameter n. Also, we see explicitly that two traversals

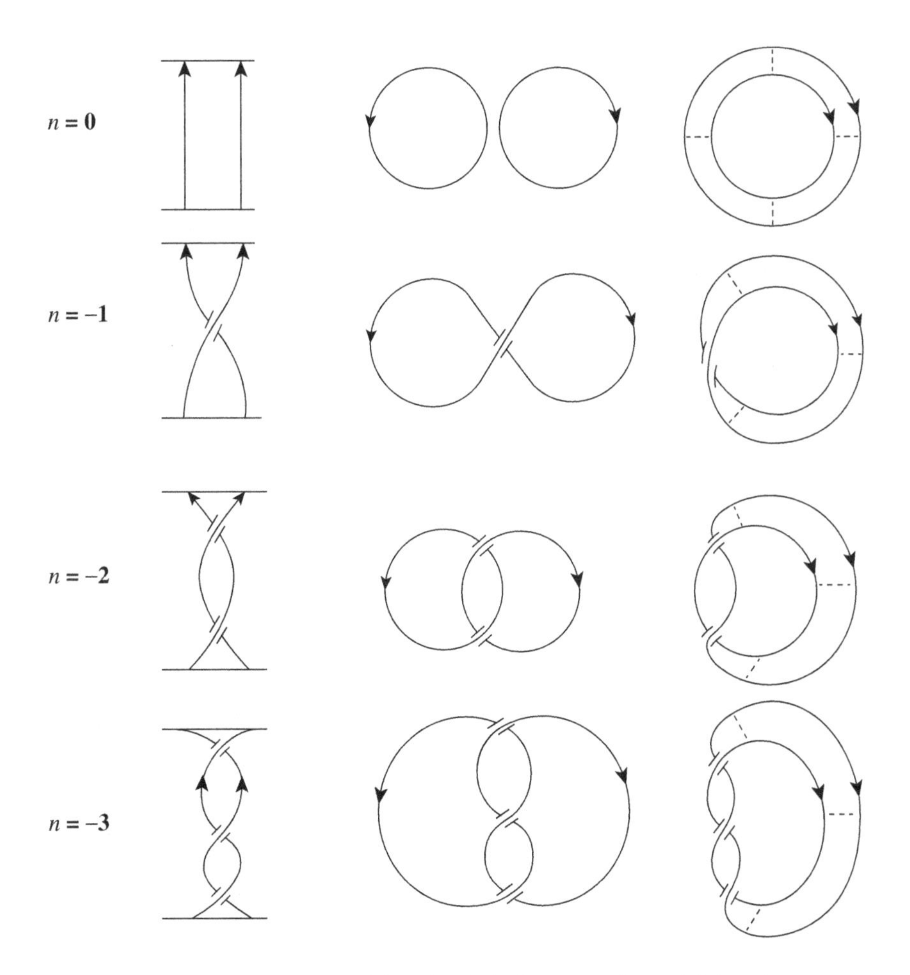

Figure 2-7. Two-strand braids with closure and partial framing.

are clearly required in the odd n case to close the knot but only one traversal for even n.

Finally, notice the dotted lines in the figure, reminiscent of "rungs" on a ladder (like the organic bases linking the two strands of DNA[1]). In fact, ***Erica Flapan***, in a fascinating book [27] on the growing relationship between Topology and Chemistry calls

[1] As per [10], our bosons (even values of NHT) evoke the "cyclic, duplex DNA — two-stranded DNA whose otherwise free ends are connected to each other".

two-strand braids with rungs and $n = \pm 1$ closure "Möbius Ladders" and shows how chemicals with that kind of configuration can be analyzed by knot theory to determine chirality (whether they are invariant to mirror reflection or not). In our case, another way to realize the aforesaid "*framing*" of a (2, n) torus knot is just to ***increase*** rung ***density*** to the continuum limit. The equality between knot parameter (n) and the MS parameter NHT is also apparent here. To quote [10], "The braided representation makes manifest that all our "particles" belong to the *same genus*, namely the set of framed (2, n) torus knots (or links)". In summary, it would appear legitimate to say that a torus knot (or link) ***is*** a two-strand braid with closure, and a Möbius strip ***is*** a framed two-strand braid with closure!

Well, for the time being it would appear that we have said all we need to about knots (and braids) even though the subject of knots and braids and their relationship to fundamental physics has evoked increasing interest in recent years. We just mentioned Flapan's book in that regard and **Louis Kauffman's** encyclopedic book [26] which has attained biblical stature (and multiple printings) therein is a must read. Also recommended is ***Alexei Sossinsky's*** small book [28] which covers a surprising amount of ground and is especially insightful. Actually, all this interest in knots is not really a completely new phenomenon, nor, in fact is their relationship to particles. In that regard we must note the primacy of another of our historical progenitors, namely William Thompson, **Lord Kelvin** who, in a sense, also anticipated our model by something like a century and a half! It is said that he got his idea after watching smoke rings, but in any event, what he came up with [4] was the notion that the various *atoms* known in that day existed as just different configurations of *knots* in the ether, the name given to the medium for the propagation of James Clerk Maxwell's electromagnetic waves. Given Kelvin's influence, the idea attracted considerable interest and some prominent adherents (including Maxwell himself) but, given the ensuing history, its demise was inevitable.

Alright, ***now*** back to our main story line about FMS and particles. As per the above, the apparent parallel between FMS structure and Gell-Mann and Zweig's ground-breaking theory of quarks, *two ups and a down for the proton* and *two downs and an up for the neutron* was not lost on me, nor, I'm sure, did it sneak by you either. As a result, the development of the alternative model then ensued in a way that seemed to have a life of its own, a phenomenon that may become evident as the model description develops in the rest of the paper. However, one thing becomes reasonably clear from the above: the "quirks" at the corners of our FMS must function at least in some respects as surrogates for quarks. That being the case, to proceed as in the SM, each quirk must be associated with an electric charge. You're probably way ahead of me here; you expect to see "up" quirks with a "charge" of $+2/3$ and "down" quirks with $-1/3$, right? However, to proceed in a more systematic fashion, note that we have six entities: four basic fermions and two quirks. Thus, assuming that fermion electric charge is simply the sum of quirk charge leads to the following set of equations:

$$\begin{aligned} 3q_d &= q_A, \\ 2q_d + q_u &= q_B, \\ q_d + 2q_u &= q_C, \\ 3q_u &= q_A. \end{aligned}$$

Prescribing a value for any two of the six charges then determines the value for the remaining four. So, for example, setting $q_B = 0$ and $q_C = +1$ does, indeed produce the right charge values for the two quirks as well as the values, $q_A = -1$ and $q_D = +2$. That's all; no big deal but we do see that charge and twist are not arbitrarily connected. Note, also, that if all charge values were in terms of the $+e$, the charge of the proton, we have

$$|q_A| = |q_C| = e,$$

which, we could interpret as saying that the electron and the proton have the ***same charge*** magnitudes (if we interpret A and C of Figure 2-7 as being the electron and proton, respectively, that is!) How about that?

Now that we have introduced electric charge into our discussion it is time to be a little more basic about the concept of an FMS.

First of all, by definition, it is ***essentially*** planar, that is to say spatially *two-dimensional*. However, since it also persists over time, we can characterize an FMS as a ***2 + 1 dimensional manifold*** that occupies a four-dimensional, $(3 + 1)$ spacetime. In other words, at this point, we are introducing what may seem like an arbitrary assumption and, it turns out, one with major consequences: in this book, the ***out-of-plane dimension*** is to be identified with ***time***. Thus, in Figure 2-5, above, showing an MS with NHT $= 3$ as the concatenation of trefoil knots ($n = 3$), the coordinate out of the plane is to be construed as the time direction. As a result, the band projects sinusoidally onto both the temporal and the spatial directions so that, for example, the three regions in the figure where the projection appears to reduce to a point are actually where the band ***extends totally in time***.

Actually, that identification is not without precedent, but all I can say about it at this time is to mention the words "**Kaluza–Klein** theory" and its extension of Relativity [29], and to promise some discussion in Sec. V. In any event, a number of important results follow, including the *manifestation* of CPT invariance and the Wheeler–Feynman notion that antiparticles can be viewed as particles moving backward in time. As we shall see, demonstrating these often complex-seeming concepts is gratifyingly straightforward, being facilitated by the restriction to a $2 + 1$ dimensional manifold. In fact, it turns out, not surprisingly, that these two results are basically manifestations of the same situation. However, at this point, another matter looms large and we need to postpone CPT demonstrations until we get into matters of Symmetry in Sec. IV for one of them and, for another, combined with the Wheeler–Feynman idea, until the more comprehensive discussion of time in Sec. VI.

The more timely (!) matter of interest alluded to amounts to what I consider to be another major *Epiphany*! To describe it we begin with the relationship we notice between the twist and the charge of the four basic fermions, namely,

$$q = (\mathrm{NHT}/2 + Q)e, \tag{2-1}$$

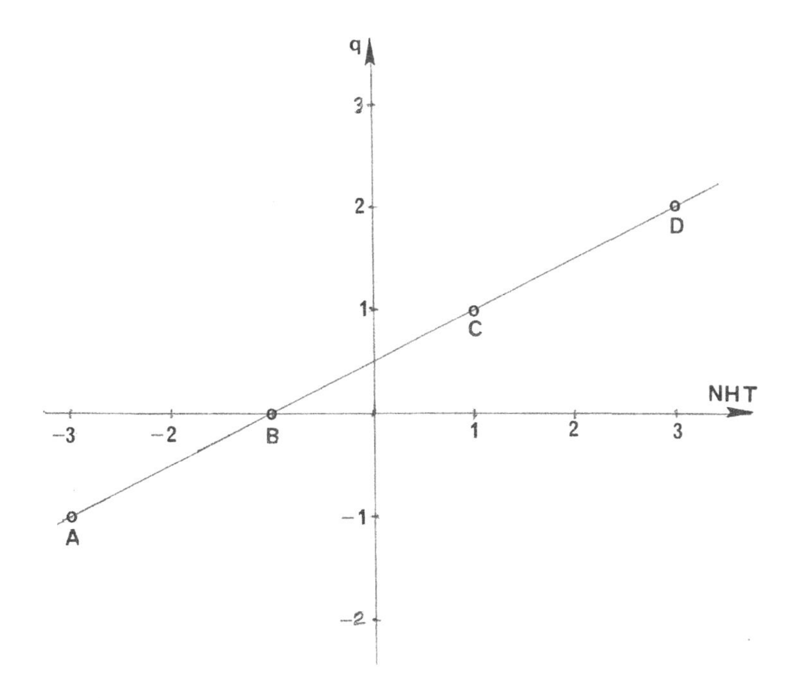

Figure 2-8. Twist vs. charge for the four basic fermions.

which we plot in Figure 2-8 (For future reference, note the ***odd symmetry***). Here q is the individual charge of any of the four, Q is the average value for all of them and e is the charge of the proton. As we see, charge ranges from $q = -e$ to $2e$ as twist ranges from NHT $= -3$ to $+3$ and $Q = e/2$.

Now there is also a well-known Standard Model relationship between ***charge*** *and* ***isospin***, namely the **Gell-Mann/Nishijima formula**,

$$q = (I_3 + Y/2)e, \tag{2-2}$$

where I_3 is the (third component of) strong ***isospin*** and Y is the so-called hypercharge which, in the case of the nucleons is $Y = B + 1$ where B is the baryon number. Thus, if we equate NHT with $2I_3$ and Q with $Y/2$ there is a *formal* ***equivalence*** between the two formulas. For example, looking ahead a bit, we have, with $Q = e/2$, and NHT$= -1$ and $+1$, for fermions B and C, respectively (also evident from the figure) that $q = 0$ and $+e$, respectively, which corresponds to equating B to the ***neutron*** and C to the ***proton***.

However, what I want to emphasize here is the identification of ***twist with isospin***, something we will rely upon throughout the rest of the book. On the one hand, ***twist*** is the *essential* ***geometrical*** *parameter* that characterizes our set of Möbius strips and associated knots. And on the other hand, ***isospin***, originally postulated by Heisenberg, has become ***the particle*** *parameter* whose elucidation in terms of Yang–Mills (non-Abelian) SU(2) gauge theory is generally characterized as the breakthrough that became the model for the consequent development of elementary particle theory in the latter half of the 20th century! Thus, we see here in an explicit, graphical way that ***geometry*** *determines* what we can say about *elementary* ***particle physics***.

We shall have more to say about the role of twist throughout the book beginning with the development of a *taxonomy* in Sec. II. In fact, looking ahead a bit, we will soon identify ***two sets*** *of three* of the four basic fermions, A, B and C for one set and D, C and B for the other, each a spin-1/2 vector space representation of SU(2), to serve as two redundant bases, existing side-by-side, each in the FYS tradition for developing a taxonomy. A similar set of "***antifermions***" is formed simply by ***reversing*** the direction of traverse, a definition that is justified in what follows on geometrical grounds in this section and on physical grounds in the next section. All quirks then become "antiquirks" — same magnitude as the corresponding quirks but opposite sign of electric charge.

In this book, although taxonomy building does not begin in earnest until the next chapter, here is the key to it: Particles with no twist as well as with all higher values of twist than three can be formed as ***composites*** by joining basic FMS and anti-FMS in an operation called ***fusion***. Fusion always joins a ***fermion*** and an ***antifermion*** in a manner that maintains continuity *of* ***traverse***. An example of (the result of) what's called ***first-order fusion*** is shown in Figure 2-9. Note the reversal of traverse in the right hand part of the figure.

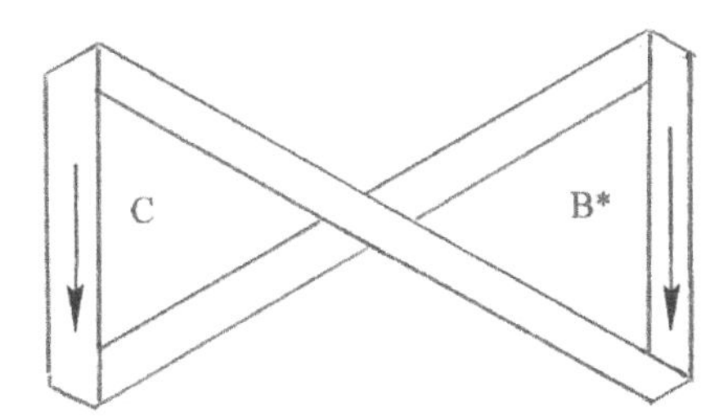

Figure 2-9. Example of first-order fusion.

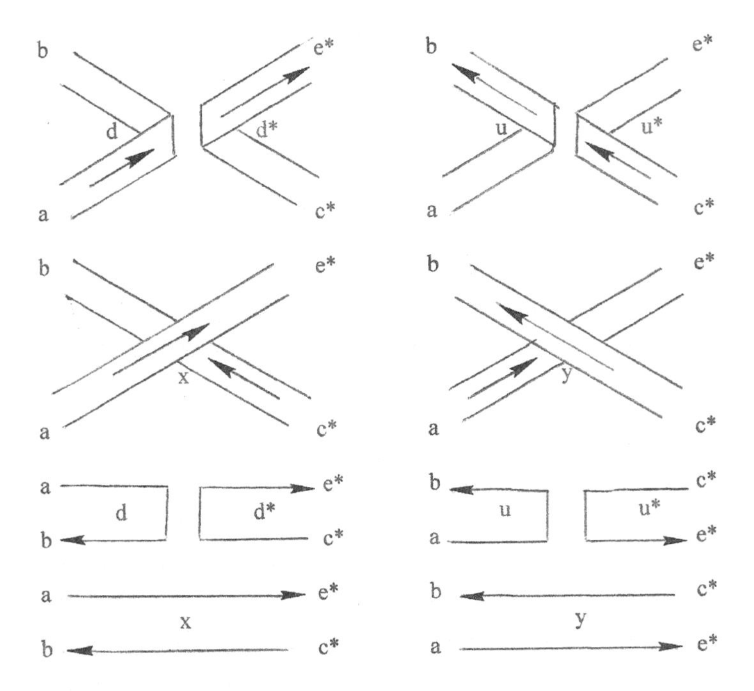

Figure 2-10. Free and fused quirks.

More specifically, it can only take place between a ***quirk*** of the fermion and a corresponding conjugate ***antiquirk*** of the antifermion. Figure 2-10 presents some detailed, operational views for both the ***free*** and the ***fused*** conditions and for both allowable situations, that is either d–d* mating (call that an **"x" junction**) or a u–u* mating (call that a **"y" junction**) with a "plan view" at the top and an "edge view at the bottom. In the ***free*** condition the edge view shows the connection being made "vertically" (out of the plane of the planform) and in the fused condition it is made horizontally. In order for ***fusion*** to occur, the vertical connections must be disenabled and the horizontal connections enabled, something we will talk about later in terms of the connection to a ***Hopf*** algebra [26, 28, 30].

All of which brings up an interesting item of mid 20th century history that we mentioned briefly in the Introduction: in 1949 **Enrico Fermi** and **C. N. Yang** wrote a paper entitled "Are mesons elementary particles" [21]. The introduction to the paper contains the following statement:

"We propose to discuss the hypothesis that the *π-meson* may not be elementary, but may be a *composite particle* formed by the association of a *nucleon and an antinucleon.* The first assumption will be, therefore, that both *an anti-proton and an anti-neutron exist,* having the same relationship to the proton and the neutron, as the electron to the positron. Although this is an assumption that goes beyond what is known experimentally, we do not view it as a very revolutionary one. We must assume, further, that between a nucleon and an anti-nucleon strong *attractive forces* exist, capable of *binding* the two particles together."[2]

So now, if, in Figure 2-9, fermion C were associated with the proton and B with the neutron, (the asterisk indicates the antiparticle), the figure would represent a ***manifestation*** of exactly what Fermi and Yang were talking about, in this case the positively charged ***pion***! Note that it has zero twist because it joins basic FMS with twists of ± 1. The only restriction, we repeat, is that, in order to maintain traverse, we can only join a fermion to an antifermion, in fact, a quirk with its conjugate antiquirk. The junction in the figure unites an "up" quirk, u, with its conjugate, u*.

The next order of fusion will add a fermion to the products of first order fusion. Although it may not be immediately apparent, examination of Figure 2-10 reveals that the traversal requirement of *each member* is shared with its partner (in effect, the detour

[2] Emphasis added.

implements the additional traversal associated with basic fermions). The result is that ***spin is additive*** for all orders of fusion such that all *Fermions* must be associated with *odd* multiples of spin-1/2 and all *Bosons* must be associated with *even* multiples of spin-1/2.

So much for a look ahead. But before we begin to talk about the taxonomy that results from what we have presented so far, it may be a good idea to clear up something missing that may have occurred to the reader: we have been blithely referring to Flattened MS on the one hand (Figures 2-1, 2-2, and 2-3) and just unflattened MS as portrayed in Figures 2-4 and 2-5 on the other. Perhaps, the following may help clarify that omission:

Consider the world of ordinary (macroscopic) experience and a real, tangible basic MS (say a belt or similar object with some reasonable elasticity). In the physical process of flattening such an object, first we would loosen it a bit to reduce its *tension* whereupon its ***twist*** is replaced by ***writhing***, that is by a set of "loopy" curves. It can be shown that writhing is energetically less demanding then twist, but in any event, as flattening continues, the curvature ends up as a set of three folds (that is to say, very small, localized half-loops) — in other words the quirks at the corners of a triangular platform (or the contiguous combination of such) as we discussed in the above. The tradeoff between twist, NHT, and writhe, W, is encapsulated by the invariance of a linear combination of the two, what is known as "linking number", namely NL = NHT/2 + W. Given a knot or link there is a simple procedure for determining linking number, which will not be given here [cf. 26] but, in the case of an FMS, we have NL = NHT/2 because all the twist disappears and is replaced by writhing, flattened into an equal number of quirks as per the above.

From a more formal point of view, we can probably all agree that the linear constraint imposed above between θ and ϕ in the torus knot representation discussed above need not apply to physical MS in real space, although θ must be *topologically quantized* over the extent of traverse in ϕ, there is no unique way to accomplish it as long as the condition

$$\int_0^{4\pi} f(\phi)d\phi = n\pi, \quad n = 0,\ \pm 1,\ \pm 2,\ \text{etc.} \tag{2-3}$$

where $(f(\phi) = d\theta/d\phi)$ is satisfied.

In the case of an FMS the change in θ just occurs in a discrete manner, namely at the quirks. In fact, again in the world of ordinary experience, an alternative to flattening a real, tangible MS is a *synthetic* approach in which an untwisted strip, ribbon, belt, etc. is *folded* at a discrete set of points before its ends are joined. Note that θ can change only by $\pm\pi$ radians at each quirk because "fold", here, means that the ribbon executes *half a revolution* about an axis

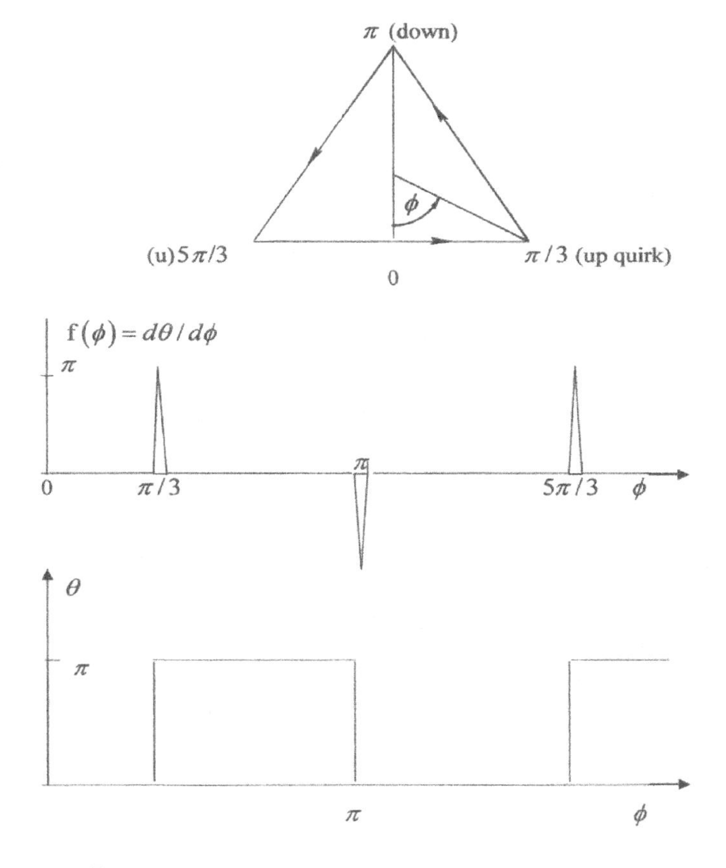

Figure 2-11. Encapsulating the synthetic approach.

in the plane of the resultant FMS (the twist axis). There is, of course, also a requirement for closure in the plane, namely

$$\sum_{1}^{N_q} \Delta\phi_i = 2\pi, \tag{2-4}$$

where $\Delta\phi_i$ is the change in the ribbon's bearing in the plane at the ith quirk and N_q is the total number of quirks. The quantum condition becomes

$$\begin{aligned} \int_0^{4\pi} f(\phi)d\phi &= \int_0^{4\pi} \varepsilon_i \pi \delta(\phi - \phi_i)d\phi \\ &= \pi \sum_{1}^{N_q} \varepsilon_i F_i = n\pi, \end{aligned} \tag{2-5}$$

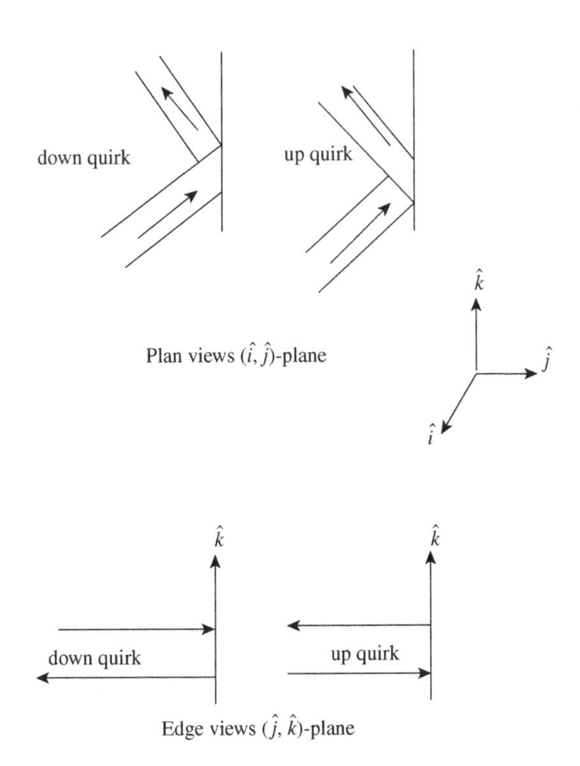

Figure 2-12. Quirks as impulsive reflections.

where $\varepsilon_i = +1$ for ccw rotation (an "up" quirk), -1 for cw rotation (a "down" quirk), and F_i is the unit step function at ϕ_i. Both requirements are shown in Figure 2-11 for a C fermion (2 ups and a down).

You can easily satisfy yourself that ***three*** folds is the ***minimum*** number ($N_q = 3$) you need to close the diagram. Figure 2-12 then shows, in idealized fashion, what is meant by an "up" quirk and a "down" quirk; each vector (directed ribbon segment) can be viewed as impulsively reflecting off a wall in the $(\hat{\imath}, \hat{\jmath})$-plane (the plane of the FMS) with the $\hat{k}$ direction normal to the plane.

So now, ***finally***, let us begin to develop a taxonomy. Meaning, to begin with that, as per FYS, we use the basic fermions we identified in the preceding to extend the list of particles we can talk about. The first extension is via first-order fusion.

3

First-Order Fusion

We begin with a sort of birds eye view by considering an abstract, *group theoretic* approach which bypasses the detail but summarizes the general architecture of the taxonomy. In this (top-level) approach (cited in [10] as per the treatment in [15]), a particle hierarchy is developed as the *direct product of* ***vector spin spaces*** *parametrized by* ***spin***. Correspondingly, the abstract result is expressed as the direct sum of subsidiary spin spaces, the so-called Clebsch–Gordan decomposition [17]. With the additional recognition that the applicable group structure (see Sec. III) is that of the ***gauge group*** SU(2), the result of the direct product of vector spin spaces with spins s_1 and s_2 is

$$V_{s_1} \otimes V_{s_2} \to V_{|s_1 - s_2|} + V_{|s_1 - s_2|+1} + \cdots + V_{s_1+s_2} \tag{3-1}$$

which equals $V_0 + V_1$ (particle spin $= 0$ or 1) for the case of first order fusion, that is, for $s_1 = s_2 = 1/2$. To begin with, we will be specifically concerned with the *entire* ***vector*** *of four* basic spin-1/2 *fermions* $V = (\mathrm{A, B, C, D})^T$ and its conjugate vector of *antifermions*, $V^* = (\mathrm{A}^*, \mathrm{B}^*, \mathrm{C}^*, \mathrm{D}^*)^T$ and the direct product

$$M = V \otimes V^{*T}, \tag{3-2}$$

the result being the (self-adjoint) matrix of 16 two-element composites shown in Eq. (3-3). In analogy with quantum mechanics, we note for future reference that M can be viewed as the ***operator***,

$$M = \begin{bmatrix} [\mathrm{AD}^*] & \mathrm{BD}^* & \mathrm{CD}^* & \mathrm{DD}^* \\ \mathrm{AC}^* & \mathrm{BC}^* & \mathrm{CC}^* & \mathrm{DC}^* \\ \mathrm{AB}^* & \mathrm{BB}^* & \mathrm{CB}^* & \mathrm{DB}^* \\ \mathrm{AA}^* & \mathrm{BA}^* & \mathrm{CA}^* & [\mathrm{DA}^*] \end{bmatrix}. \tag{3-3}$$

Note the two sets of three fermions each, (A, B, C and D, C, B), alluded to above, in the grouping of six matrix elements in the lower left-hand corner and the similar grouping of another six elements in the upper right-hand corner. Notice also the two elements in *square* brackets in the corners of the matrix; in reality, these elements ***cannot exist*** because (with reference to Figure 2-3) no common quirk–antiquirk combination exists in either case. Also, there are actually *only* ***two*** V_0 elements, namely, CB*, the one shown above in Figure 2-15, and its conjugate, BC*, shown in Figure 3-1.

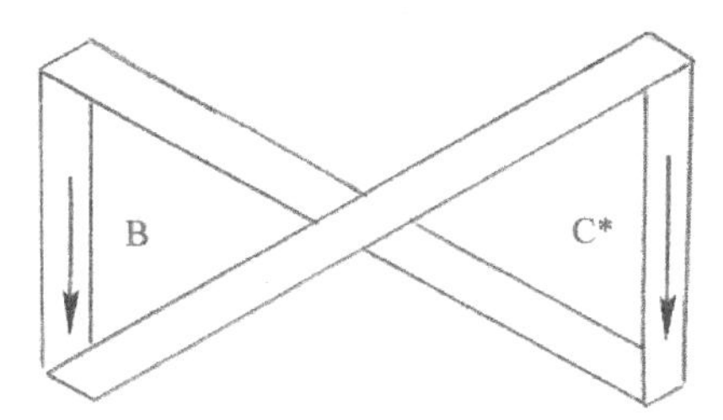

Figure 3-1.

Although, as per the above, these two elements can be formed by fusion, topologically, they are both just *doubled-over versions* of the trivial, zero-twist MS. In other words, they are ***excited states*** of the basic ***untwisted*** state and, in fact (we recall), in each case the algebraic sum of the twists of the fused constituents is zero. The other 12 bosons are all V_1 vector bosons in their ground state and can also be formed either by fusion or directly by a twist whose NHT is also the sum of those of its constituents.

Of course, we are not forgetting that the actual, ***physical*** *process*, corresponding to the direct product, is ***fusion*** beginning with the basic set of spin-1/2 fermions. This process can be formalized in terms of the contraction of a tensor that describes the coexistence of a fermion and an antifermion in close proximity and will be discussed later on when we consider the relationship of our model and the Hopf algebra (see Sec. III).

Note that ***twist*** increases from left to right in the matrix on the R.H.S. of Eq. (3-3) for the fermion components and upward for the antifermion components. Thus the loci for composite twist are lines parallel to the principle diagonal as we indicate in Figure 3-2 and the ***gradient*** is directed along the orthogonal diagonal. Each dot in the figure represents a two-letter *word* (boson) whose associated *twist is the sum of the indicated abscissa and ordinate values.* As indicated, the loci are antisymmetric about the principle diagonal. Again, for future reference we note that the gradient of *charge* is directed downward and to the right, *normal* to the twist gradient, and the associated charge loci are antisymmetric about the orthogonal diagonal. In summary: of the product terms, there are six with positive twist, six with negative twist and four with no twist, two of which are really nonexistent and of the remaining pair, one has positive charge and the other negative charge. Note that the last two entries may be viewed as excited states of the unknot, a notion we will see again in Chap. 5.

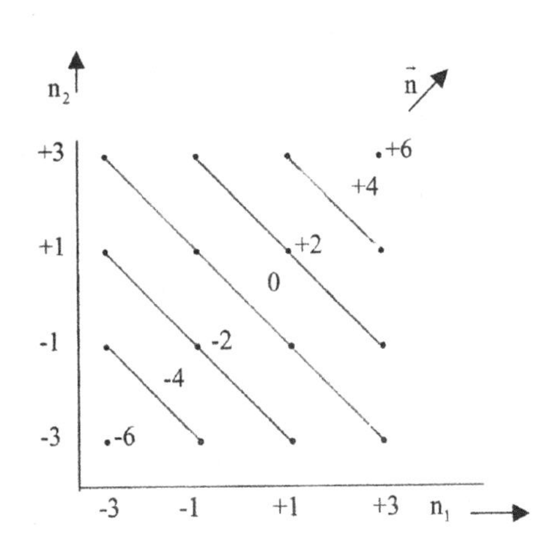

Figure 3-2. Twist loci in first-order fusion.

Note: Here and in the sequel we will use the ***letter*** **(n)** to denote twist ***instead of*** **NHT**, in keeping with the relationship between MS and $(2, n)$ torus knots as discussed above.

The reason we are focusing on *twist* is twofold: its relationship to *isospin* as noted above and its relationship to the topic of ***degeneracy*** which will be the determinant for the next lower taxonomical level after spin. Note that the notion of degeneracy is customarily associated with the invariance of a fundamental attribute shared by a multiplicity of otherwise-distinguishable versions of a physical system. In what follows, that attribute is *twist.* The structure of the associated degeneracy can then be described in terms of *combination, permutation, composition* and *contingency,* defined as follows:

- Briefly, more than one ***combination*** of integers can add up to a specified value of *twist* at a given order of fusion.
- A given *combination* of integers can be ***ordered*** in more than one way to generate a ***permutation***.
- A given *permutation* can be realized in more than one way as a consequence of the detailed ***composition*** of the constituents; and finally:
- The possibilities for a given higher order of fusion hinge on the nature of the previous fusion — thus, ***contingency***.

Composition and contingency require a more detailed consideration and will be so discussed in Chap. 8. However, aggregation in terms of the degeneracy *of twist* for first order fusion is rather simple and can be accomplished simply by ***inspection*** — that is by assembling the entries along each diagonal locus in the R.H.S. matrix of Eq. (3-3). However, noting that those slanted loci illustrate the relationship, $n_2 = n - n_1$, we are led to the formal notion of ***symbolic convolution***; where the direct product matrix M presents the requisite information *visually,* convolution does the same thing ***algebraically***. For the case of first-order fusion, we therefore define the operation of (symbolic) convolution to be

$$\gamma_j = \sum_{i=0}^{j-1} \beta_{i+1}\alpha_{j-i}, \tag{3-4}$$

where

$\alpha_\mu = \mathrm{A, B, C}$ and D for $\mu = 1, 2, 3$ and 4, respectively,
$\beta_\nu = \mathrm{A^*, B^*, C^*}$ and $\mathrm{D^*}$ for $\nu = 1, 2, 3$ and 4, respectively, and
$\gamma_\lambda = -6, -4, -2, 0, +2, +4$ and $+6$ for $\lambda = 1, 2, 3, 4, 5, 6$ and 7 respectively.

and α_μ, β_ν and γ_λ are to be identified with the set of four basic letters, the set of four basic conjugate letters and the twist-valued headings, n_ν, in the output of the convolution operation, respectively. It helps to picture what's happening operationally as shown in Figure 3-3.

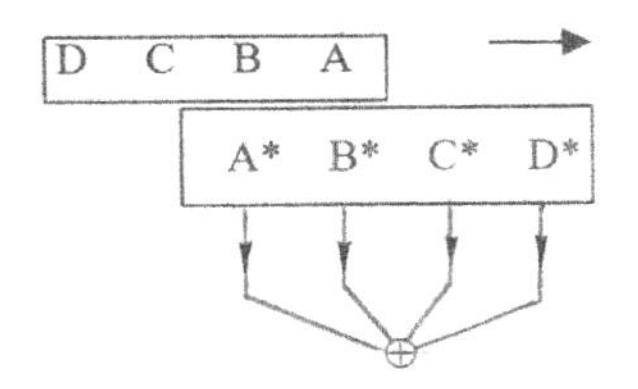

Figure 3-3. Symbolic convolution.

Of course, the summation indicated in Eq. (3-4) really denotes ***assembly*** of *products* rather than numerical summation as we see in the results of convolution shown in Figure 3-4. For example, the device in Figure 3-3 is poised to assemble AB* and BA*, the doublet with a twist of $n = -4$ as we see in Figure 3-4.

n = –6	–4	–2	0	+2	+4	+6
AA*	AB*	AC*	[AD*]	CC*		
	BA*	CA*	[DA*]	DB*	DC*	
		BB*	BC*	BD*	CD*	DD*
			CB*			

Figure 3-4. First-order fusion; twist and charge sub-groups.

Also, note the bilateral ***mirror*** symmetry with the replacement of letters A and B by D and C, respectively; this is essentially a perpetuation of the antisymmetry noted back in Figure 2-3 for the *basic* fermions. From a slightly different point of view, the figure is a non-overlapping combination of the product of each of the two three-member vector spaces (A, B and C) and (D, B and C), with its conjugate vector (A^*, B^* and C^*) and (D^*, C^* and B^*), respectively, such that the two products are mirror images of each other. Correspondingly, the sum of twist over the figure is zero and charge, which is different for each entry in a twist assembly (i.e. column) sums to zero in each twist assembly. Finally, for future reference, we note that there is a total of 14 two-letter "words", (not including the two unrealizable ones) which may be trivially viewed as assembled into permutation groups, in this case four singlets and five doublets, the latter occurring in charge conjugate pairs.

4

Dirac and Traverse

That traverse proceeds ***one way*** through the fermion and the ***opposite*** *way* through the antifermion components of the *figure-eight* configuration of our boson model highlights how yet another historically iconic item, in this case the **Dirac** theory [17], has bearing on our alternative model. According to all accounts, the notion that antiparticles exist emerges in the combination of Quantum Mechanics and Special Relativity via the Dirac theory. To set the stage for how that theory relates to our notion of ***fusion***, as implied by that theory, we *abstract* and *compress* it as follows: As is well known, the equation for a particle with rest mass m and spin-1/2 is succinctly stated by the Dirac equation

$$D\psi = 0,$$

where ψ is the quantum mechanical state vector, D is the Dirac operator given by

$$D = i\gamma^{\mu}\partial_{\mu} - m, \tag{4-1}$$

and summation over $\mu = 0$ to 3 is implied with 0, and 1, 2 and 3 corresponding to the time t, and the spatial variables x, y and z, respectively. Also, a factor of $\hbar = h/2\pi$ in the first term has been set equal to 1. Dirac imposed the SR compatibility constraint by demanding that

$$-DD^{*}\psi = (\gamma^{\mu}\gamma^{\nu}\partial_{\mu}\partial_{\nu} + m^2)\psi = 0 \tag{4-2}$$

be equal to the Klein–Gordon equation,

$$(\partial^{\mu}\partial_{\mu} + m^2)\psi = 0, \tag{4-3}$$

the quantum mechanical equivalent of the Lorentz invariant

$$E^2 - p^2 = m^2 \tag{4-4}$$

with E and p being the energy and momentum associated with relativistic bosons and

$$D^* = (-i\gamma^\mu \partial_\mu - m) = -(i\gamma^\mu \partial_\mu + m) \tag{4-5}$$

the complex conjugate to the expression in Eq. (4-1). All of which is of course well-known as is the resulting requirement that the gammas must be constant 4×4 matrices that conform to the definition of a ***Clifford algebra*** [31] (see Sec. III). This leads (*almost!*) directly to the re-expression of the Dirac equation as two, *coupled*, vector eigenvalue equations, namely

$$\begin{aligned} \partial_t \psi_\alpha + S\psi_\beta &= m\psi_\alpha, \\ \partial_t \psi_\beta + S\psi_\alpha &= -m\psi_\beta, \end{aligned} \tag{4-6}$$

where $\psi_\alpha = (\psi_1, \psi_2)^T, \psi_\beta = (\psi_3, \psi_4)^T$ and

$$S = \begin{pmatrix} i\partial_z & (i\partial_x + \partial_y) \\ (i\partial_x - \partial_y) & -i\partial_z \end{pmatrix} \tag{4-7}$$

which has traveling wave solutions with exponents proportional to linear combinations of the time and space variables, i.e. to $\omega t \pm \mathbf{k} \cdot \mathbf{x}$ where $\mathbf{k} = (k_x, k_y, k_z)^T$, ω and $\mathbf{k}$ being the radial frequency and momentum eigenvector, respectively, thus implying the ***equivalent*** *form*

$$\begin{aligned} \omega\psi_\alpha + K\psi_\beta &= m\psi_\alpha, \\ \omega\psi_\beta + K\psi_\alpha &= -m\psi_\beta, \end{aligned} \tag{4-8}$$

where

$$K = \begin{pmatrix} k_z & (k_x - ik_y) \\ (k_x + ik_y) & -k_z \end{pmatrix}.$$

Either of the last two equations expresses the operation of what amounts to a ***spin matrix***, (e.g. [31] and Sec. III) on a ***four-vector*** to generate the direct sum of two ***coupled spinors***, one with positive and one with negative energy. Note that the two interact with each other in a circularly referential fashion and in that regard, they express a ***relationship*** between the Dirac theory and our alternative particle model; that is they are talking about the same kind of situation. To see that, we begin by rewriting Eq. (4-8) as

$$\begin{aligned}\psi_\beta &= -K^{-1}(\omega - \mathrm{m})\psi_\alpha, \\ \psi_\alpha &= -K^{-1}(\omega + m)\psi_\beta.\end{aligned} \tag{4-9}$$

We can then demonstrate the ***circular*** nature of this pair of equations explicitly by the operational diagram of Figure 4-1, below, where the operators are the annotated boxes.

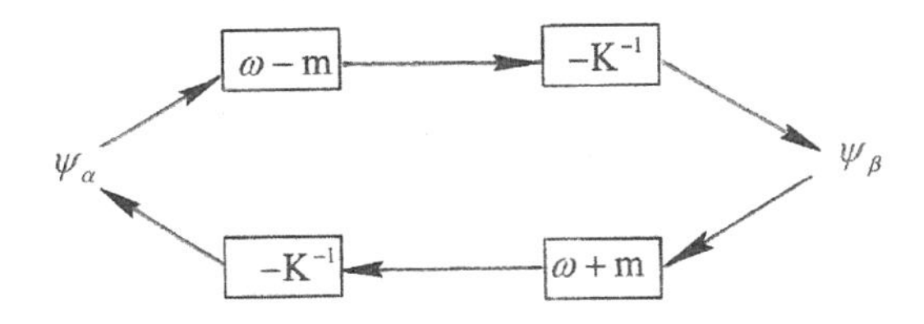

Figure 4-1. "Circularity" of the Dirac equations.

However, Figure 4-2 shows ***another*** way to display the same situation:

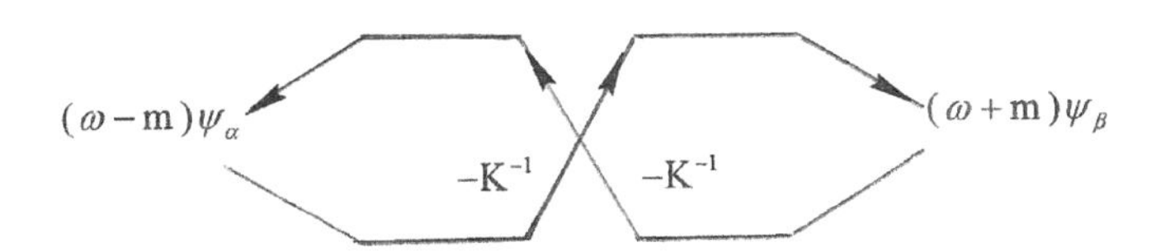

Figure 4-2. Analogy to the FMS boson relationship.

In view of the preceding discussion of fusion in this paper, the topology of this diagram is seen to be identical to that of the MS model of a boson: a bound state comprised of a spin-1/2 "***fermion***" on the *left* with *counterclockwise* traverse and *positive* mass, and its conjugate "***antifermion***" on the right with *clockwise* traverse and *negative* mass. Or, to put it another way: the MS model is also a ***manifestation*** of this emergent Dirac formalism. Which is not really surprising since the Dirac theory was really formulated in exactly this way, i.e. as a factorizable version of the Klein–Gordon equation, the two factors representing ***positive*** and ***negative*** energy situations, respectively (see Eqs. (4-2) and (4-6)), something we shall discuss in more detail when we get into the solitonic solution of some Differential Geometry in Sec. V.

5

Second-Order Fusion

As per the comment made above in connection with Eq. (2-3), we can view *second-order* fusion as *Matrix M* operating on basic *fermion vector V*, that is, in analogy with Eq. (2-1), we can write

$$M_{s_1} \otimes V_{s_2} \rightarrow P_{|s_1-s_2|} + P_{|s_1-s_2|+1} + \cdots + P_{s_1+s_2} \qquad (5\text{-}1)$$

which equals $P_{1/2} + P_{3/2}$ (spins of 1/2 or 3/2) for $s_1 = 1$ (or 0) and $s_2 = 1/2$. The result can be viewed as a 4-vector,

$$P = (P_1, P_2, P_3, P_4)^T, \qquad (5\text{-}2)$$

whose elements are the ***matrices***

$$P_k = MV_k, \quad \text{with } k = \text{A, B, C or D}. \qquad (5\text{-}3)$$

For example, P_1 is a *matrix* whose elements are *three letter words*, formed by appending the letter, A, to each of the elements of matrix, M. Geometrically, we can picture the vector, P, as a vertical stack of four horizontal planes, the P_k.

However, since we are preoccupied with the aggregation of elements in terms of ***twist***, we note that the ***twist loci***, as per the invariance of twist in fusion (see below), appear as a set of ***inclined*** *planes* where, as indicated in Figure 5-1 the values of twist range antisymmetrically from −9 to +9, a situation, in three dimensions, quite analogous to the inclined loci, in two dimensions, of Figure 3-2 for the case of first-order fusion (see below). Actually, an equivalent set of inclined planes (shown in [12] but not here) exists for *charge* and again, for future reference, the associated *charge* gradient is normal to that for twist (a circumstance we will find useful in Sec. III) and there is a corresponding set of inclined planar charge loci. The two sets of planar loci therefore intersect this time in a set of lines which are discussed in more detail below.

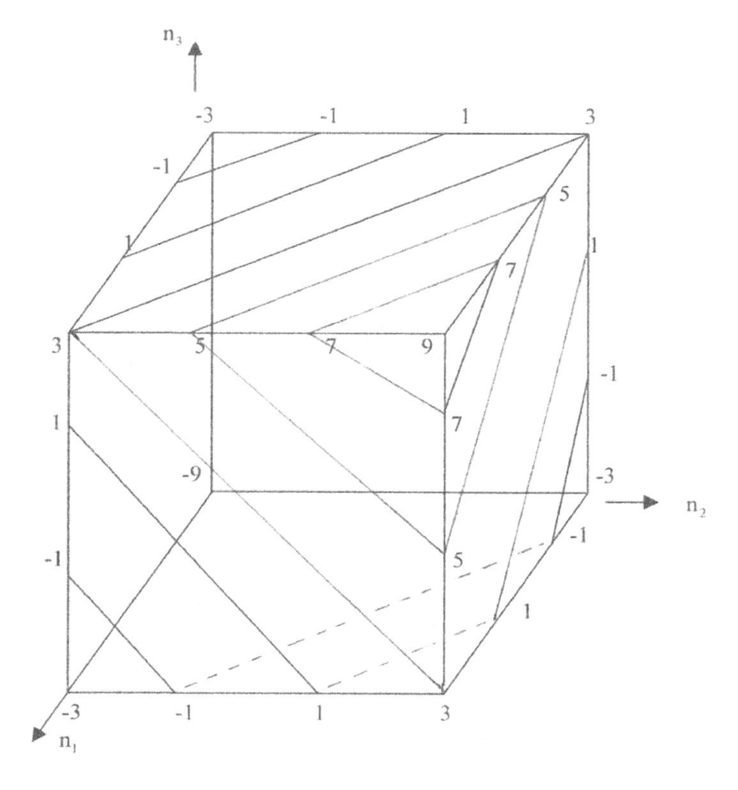

Figure 5-1. Inclined twist loci in second-order fusion.

As was the case in first-order fusion, the behavior shown in this figure is then also the result of the ***invariance of twist*** *in fusion* — in this case, we have

$$n = n_1 + n_2 + n_3. \tag{5-4}$$

Consequently, we can *again* generate the occupancy of the individual twist headings by a process of symbolic ***convolution***, where we define

$$\alpha_\mu = \text{A, B, C and D} \quad \text{for } \mu = 1, 2, 3 \text{ and } 4\text{, respectively,}$$
$$\beta_\nu = (-6), (-4), (-2), (0), (+2), (+4) \text{ and } (+6), \quad \text{for}$$

$$\nu = \ 1, 2, 3, 4, 5, 6 \text{ and } 7, \text{ respectively, and}$$
$$\gamma_\lambda = -9, -7, -5, -3, -1, +1, +3, +5, +7, \text{ and } +9 \quad \text{for}$$
$$\lambda = 1,\ 2, \ldots, 9 \text{ and } 10, \text{ respectively,}$$

such that, this time, α_μ, β_ν and γ_λ are to be identified with the set of four basic letters, the set of seven ***columns*** of Figure 2-4 and the output, respectively. An operational diagram, Figure 5-2, again illustrates the procedure, whose output is shown as the set of ten planes of Figure 5-3, which run from $n = -9$ to $+9$ with increments of $\Delta n = 2$.

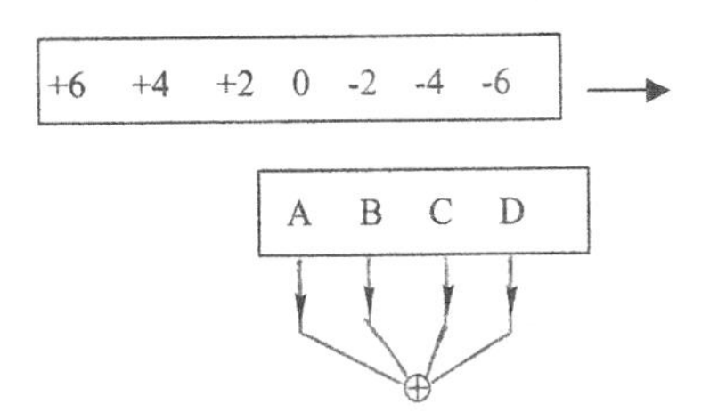

Figure 5-2. Operational diagram for second-order convolution.

The linear ***intersections*** of twist and charge loci are readily visible, in analogy with the first-order loci of Figure 3-2, as ***inclined groupings*** of charge in each of the twist loci of Figure 5-3. For

$n = -9$:	AA*A	$n = +9$:	DD*D
$n = -7$:	BA*A AB*A AA*B	$n = +7$:	DC*D CD*D DD*C
$n = -5$:	CA*A BB*A AC*A BA*B AB*B AA*C	$n = +5$:	DB*D CC*D BD*D DC*C CD*C DD*B
$n = -3$:	DA*A CB*A BC*A AD*A CA*B BB*B AC*B BA*C AB*C AA*D	$n = +3$:	DA*D CB*D BC*D AD*D DB*C CC*C BD*C DC*B CD*B DD*A
$n = -1$:	DB*A CC*A BD*A DA*B CB*B BC*B AD*B CA*C BB*C C AC*C BA*D AB*D	$n = +1$:	CA*D BB*D AC*D DA*C CB*C BC*C AD*C DB*B CC*B BD*B DC*A CD*A

Figure 5-3. Second-order fusion twist and charge groupings.

example, for $n = -7$ we have the doublet BA*A and AA*B with charge $q = 0$ and the singlet AB*A with charge $q = -2$ while for $n = -5$ we have the triplet CA*A, BA*B and AA*C with $q = +1$, the doublet BB*A and AB*B with $q = -1$ and the singlet AC*A with $q = -3$. The charge increment between loci in each case is $\Delta q = -2$. Note that 14 of the entries in the figure are really nonexistent since they juxtapose D and A* or A and D*.

Figure 5-4 then breaks out the charge groupings explicitly. Actually, there is no algebraic reason not to use charge rather than twist as the determinant of spin degeneracy. However, since the particle attribute of *isospin* is associated with the basic *physical* parameter of *twist* while discrete quirk *charge* emerges, by definition, upon flattening we consider twist to be the more fundamental.

Finally, in analogy with Figure 3-4 for first-order fusion, Figure 5-5, below, summarizes all the results of convolution. And

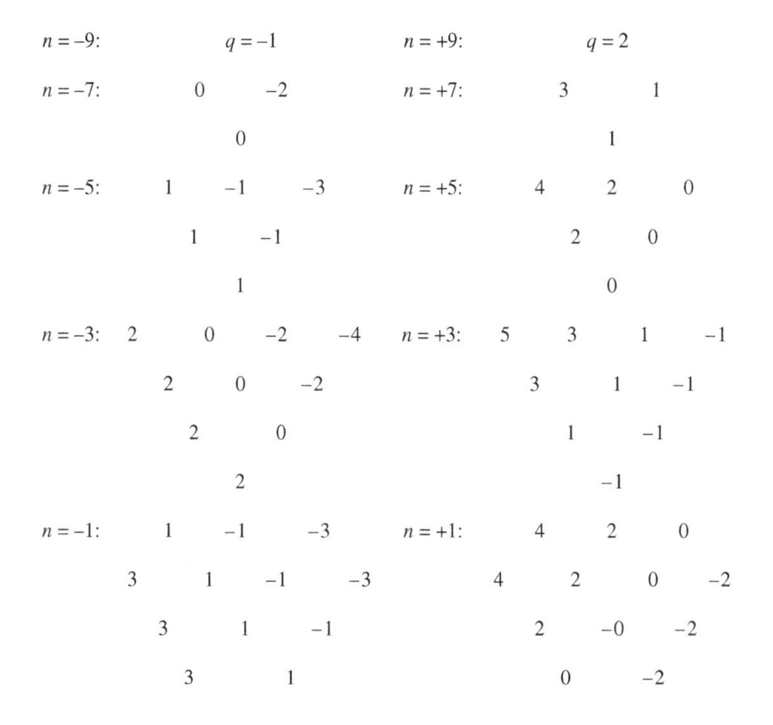

Figure 5-4. Second-order charge grouping in detail.

NHT =	−9	−7	−5	−3	−1
	AA*A	AA*B AB*A BA*A	AB*B BA*B BB*A AA*C [AC*A] CA*A	AB*C CA*B BC*A CB*A AC*B BA*C BB*B [AA*D] [AD*A] [DA*A]	AC*C CA*C CC*A AB*D [DA*B] [BD*A] DB*A [AD*B] [BA*D] BB*C BC*B CB*B

NHT =	9	7	5	3	1
	DD*D	DD*C DC*D CD*D	DC*C CD*C CC*D DD*B [DB*D] BD*D	DC*B BD*C CB*D BC*D DB*C CD*B CC*C [DD*A] [DA*D] [AD*D]	DB*B BD*B BB*D DC*A [AD*C] [CA*D] AC*D [DA*C] [CD*A] CC*B CB*C BC*C

Figure 5-5. Summary of second-order fusion groupings modeled by convolution.

again, we note that the entries of that figure under the $NHT = \pm 1$ and ± 3 headings, are ***excited states*** of the basic spin $= 1/2$ fermion configurations. However, all the other entries are in the minimal spin $= 3/2$ state. Also note the carry-through of the original *antisymmetry* and division into two groups of basic fermions discussed above, here displayed vertically due to lateral space limitations. Again for future reference (see below) there are a total of 64 permutation groups, 4 singlets, 12 triplets and 4 sextets, in each half, again including the unrealizble entries, and, again we note that every entry in the upper half of this figure has a counterpart in the lower half.

Actually, excited states can be created, not only by *fusing* the indicated component fermions, but also by ***additional folding*** of the fermionic states subject to excitation. This folding process is equivalent to a what is known in Knot Theory as a Reidemeister move [26]. It preserves the ***twist*** but not the ***charge*** of the original unexcited fermion. Both of these processes will be seen to be involved in the interactions modeled in Sec. IV, the Standard Model Connection.

6

Combinatorial Perspective (Analysis)

The preceding has been quite general in that we have kept explicit mention of the Standard Model to a minimum (for a reason that shall be divulged in Sec. IV). In fact, we could just as well have been talking about the ***lexicon*** of a simple language in which the ***words*** are limited to one, two or three ***syllables***. The syllables are the four basic fermions A, B, C and D and their conjugate counterparts, each spelled out in terms of three *letters*, drawn from a two-letter alphabet consisting of the letter u (pronounced "up"!) and d (pronounced "down") and their conjugate counterparts. As we have seen, the language is so simple that there is only one constraint in its development, namely that juxtaposition (called "fusion" in the preceding) can only take place between a syllable and a conjugate syllable (not necessarily its own conjugate) which implies that letter d can fuse only with d* and u only with u*. Statements in this language are then to be ***translated*** into the language of the SM, a task we accomplish in Sec. IV.

Also as we have seen, ***complexity*** in this picture, arises in the form of ***degeneracy*** as a result of the detailed ***spelling*** of the syllables and their triangular constitution which allows them to combine in more than one way. The results obtained to this point, which were summarized in Figure 3-4 for first order and Figures 5-3 and 5-5 for second order fusion, considered only the effects of combination are then readily ***predictable*** on a ***combinatorial basis*** and we finish this section with a short discussion to that effect. To begin with, we shall refer to the basic fermions as the letters A, B, C and D for which the corresponding NHT $= -3, -1, +1$ and $+3$, a contiguous, balanced sequence of odd integers. First, we note that the arithmetic of fusing two basic letters (the second a conjugate letter) taken independently from the basic letter list means that we must end up with a contiguous, balanced sequence of even integers: seven values of *twist*, namely NHT $= 0, \pm 2, \pm 4$, and ± 6. Similarly, fusing three letters together (with the middle one a conjugate letter) produces a contiguous, balanced sequence of odd integers: ten values of twist, namely NHT $= \pm 1, \pm 3, \pm 5, \pm 7$ and ± 9.

Call the ***assembly***, at each order of fusion, OF, of the ***permutations*** for a given value of twist, a *twist* assembly, TA. That is, the permutations are organized into permutation groups, distinguished from one another by their *twist* and the permutation groups are *combined* by virtue of their common order of fusion. There are T such assemblies (because there are T values of twist) with, as we have seen, $T = 7$ at OF $= 1$ and 10 at OF $= 2$ or, in general, $T = 3$ OF $+ 4$. As a collection of TAs, each OF has a complete set of permutation groups — that is a complete set of *combinations*. To show this explicitly, we define

L = the number of basic *letters* (or integers) available for combination,
W = the number of letters in a *word* at a given OF,
S = the number of *identical* letters per word,
C = the number of *combinations* for a given set of L, W and S values.

(Here, we have used $L = 4$, and $W = 2$ and 3 for the first and second OF, respectively). For the case of $S = 0$ (all letters are different), the number of combinations is just

$$C = L!/W!(L - W)! \tag{6-1}$$

which gives $C = 6$ and $C = 4$ for the cases of $W = 2$ and 3, respectively (the first and second OF, respectively). The members of each permutation group are distinguished from one another by their *charge*. The number of permutations in a permutation group also depends on how many letters are different. For the case of all letters are different, permutation of each of the combinations then implies that we have six doublets (two permutations) and four sextets (six permutations), for the first and second OF, respectively.

For the rest of the range of duplication of letters, $0 < S \leq W$, we can write

$$\begin{aligned} C &= L(L-1)!/(W-S)![(L-1)-(W-S)]! \\ &= L!/(W-S)!/[(L+S)-(W+1)]! \end{aligned} \qquad (6\text{-}2)$$

which gives $C = 4$ for the case of $S = W$ (all letters are the same) for both $W = 2$ and 3, i.e. for both the first and second OF. These are singlets, one word for each letter of L, in both cases. Finally, for the intermediate case of $W = 3$ and $S = 2$ (a triplet for the second OF) we find $C = 12$. In summary, we have computed four singlets and six doublets for first-order fusion for a total of 16 words, and 4 singlets, 12 triplets and 4 sextets for a total of 64 words for second-order fusion which matches what we see in the figures. And finally, a caveat mentioned in the foregoing: permutations cannot actually exist in cases such that letter A fuses with letter D* or letter D fuses with letter A*.

7

More Perspective

In summary to this point, we reiterate: our Alternative Model has promoted the status of the MS/knot genus from that of a device used to *demonstrate* the phenomenon of spin and the gauge group SU(2) to the ontological *basis* for a particle model. But from an even more fundamentally ontological level, you may recall the comment in the Introduction to the effect that underlying the attributes that characterize our particles is their ***toroidal*** topology. Thus, broadly speaking, ***spin*** (more accurately, order of fusion) serves as a ***top-level***, SU(2) *discriminant* for organizing FMS taxonomy. We have seen that two traversals of the MS are required for *all odd* values of its NHT (which we recall is equal to the value of the n of the associated knots) but only one traversal for all *even* NHT and that ***higher values*** of NHT assume the form of ***composites*** of triangular planforms. In such composites the traversal requirement of each member is now shared with its partners (in effect, the detour implements the second traversal requirement of a single FMS). Note that each member of a composite still retains its ***toroidal*** topology so that, in essence, the overall taxonomy described above consists of tori of genus one, two and three which is to say that we are concerned with tori with ***one, two and three holes***!

In any event the result is that ***spin is additive*** such that MS with *odd* values of NHT (alternative model *fermions*) are assigned corresponding *odd* multiples of spin-1/2 and those with *even* values of NHT (alternative model *bosons*) are assigned corresponding *even* multiples of spin-1/2. We thus have the ***basic fermions*** with spin-1/2; the ***first-order*** *fusion* ***bosons*** with spins 0 and 1;

and the ***second-order*** fusion ***fermions*** with spins-1/2 and 3/2). Consequently, the number of spin-1/2 multiples in a given composite (either 0, 2 or 3) is the same as the number of its basic triangular planforms *except* where the composite is actually an excited state of a lower order of spin.

As we have seen, ***twist*** (NHT) in this model is identified with ***isospin*** and thus serves as an organizing *principle* for the ***degeneracy*** of the spin categories as discussed and illustrated in the preceding in considerable detail. One way to *encapsulate* taxonomical organization down to this level (exclusive of the effects of composition, permutation and contingency) is in terms of a twist-based, ***quaternary*** number system. As background, we recall the long-known "Belt Trick" (see [26] for full graphic demonstration!) in which an unbuckled ***belt*** with two half-twists (one full twist) whose ends are held cannot be untwisted whereas one with four half twists (two full twists) can! Thus, in a sense, four half-twists and no twist are equivalent. We note in passing that the BT is sometimes cited as a way to provide some physical manifestation of what's involved in the somewhat mysterious phenomenon of spin.

More generally, any number of half-twists (NHT) is equivalent to NHT +4 or NHT −4 depending on the direction in which the trick is performed. Consequently, the BT can be associated with the ***quaternary*** number system, the group of integers under addition modulo 4, a particular manifestation of the ***cyclic group*** *of order four* [15]. One way to characterize that group is as an array of four rows a, b, c and e of numbers such that $ab = c$, $ac = bb = e$, and $ae = ea$, etc., where the group operation is addition modulo four and e is the "identity row". If we also associate such an arrangement with the NHT of a ***closed*** belt with positive and negative integers signifying twist to the right and left, respectively, it is convenient to write the array as follows:

c:	−9	−5	{−1}	{3}	7	11	15	etc.
b:	−10	−6	−2	2	6	10	14	etc.
a:	−11	−7	{−3}	{1}	5	9	13	etc.
e:	−12	−8	−4	0	4	8	12	etc.

With the understanding that there is a duplicate conjugate array, we can then associate the set of four numbers in ***brackets*** with the set of four basic *fermions*, the set of seven underlined ***even*** integers with two-letter *bosons* and, finally, ***all*** the ***underlined odd*** integers as the set of one-letter plus three-letter fermions. Thus the overall set of underlined numbers includes all the FMS discussed in the foregoing.

The magic number, 4, also shows up in the Gell-Mann/Nishijima relationship associated with the four basic fermionic particles. We restate it here in simplified form as

$$q = (n + 1)/2, \tag{7-1}$$

where n = NHT, the MS half twist. Now consider the quadratic polynomial

$$Q(q) = q^2 - 1 = 0 \tag{7-2}$$

its roots are $r_q = -1$ and $+1$ which we tentatively identify with the ***charge*** of particles A and C. However, suppose we *transform* the polynomial of Eq. (7-2) as per Eq. (7-1). The result is

$$N(n) = (n^2 + 2n - 1)/4 = 0 \tag{7-3}$$

whose roots are $r_n = -3$ and $+1$ which are indeed the ***twists*** of particles A and C. A similar relationship is readily found to hold for the twist and charge relationships associated with particles B and D and their twists and charges. All of which further emphasizes the (mathematically simple) *topological* origin of the G-M/N relationship.

For composite particles, although the situation is more complex because of the degeneracy, we can use Eq. (7-1) to compute twist and charge as

$$n = n_1 + n_2 \quad \text{and} \quad q = q_1 - q_2 = (n_1 - n_2)/2 \tag{7-4}$$

for first-order fusion and

$$n = n_1 + n_2 + n_3 \quad \text{and} \quad q = q_1 - q_2 + q_3 = [(n_1 - n_2 + n_3) + 1]/2 \tag{7-5}$$

for second-order fusion. We can then extend the discussion by expressing the relationship in *invariant* form by defining

$$N = n + 1 - 2q. \tag{7-6}$$

For basic fermions we must have $N = 0$ by definition, but for first-order fusion we can use Eq. (7-4) to compute

$$N_1 = 2n_2 + 1 \tag{7-7}$$

independent of n_1, which, in terms of the matrix M, defined above, translates to a matrix with four identical rows, and columns that increase by a numerical increment of ***four*** per column,

$$N_M = \begin{vmatrix} -5 & -1 & 3 & 7 \\ -5 & -1 & 3 & 7 \\ -5 & -1 & 3 & 7 \\ -5 & -1 & 3 & 7 \end{vmatrix}. \tag{7-8}$$

Similarly, for second-order fusion, this computes to

$$N_2 = 2n_2 \tag{7-9}$$

which translates to

$$N_P = \begin{vmatrix} -6 & -2 & 2 & 6 \\ -6 & -2 & 2 & 6 \\ -6 & -2 & 2 & 6 \\ -6 & -2 & 2 & 6 \end{vmatrix}, \tag{7-10}$$

with identical rows and with columns also increasing by a numerical increment of *four*, for each of the four planes, $P1$ to $P4$ of the three-dimensional array discussed, in Chap. 4.

Well, so much for numerology! Some additional perspective on model development as discussed so far is what might be termed the taxonomical ***information*** content associated with the *combinatorics* as summarized in Figures 3-4 and 5-5. That is, we can use the total number of entries, say S, in each twist column of the figures to compute a corresponding set of "bits", say $B = S$ Log S, with the result portrayed in Figure 7-1 where 0 and x indicate first- and

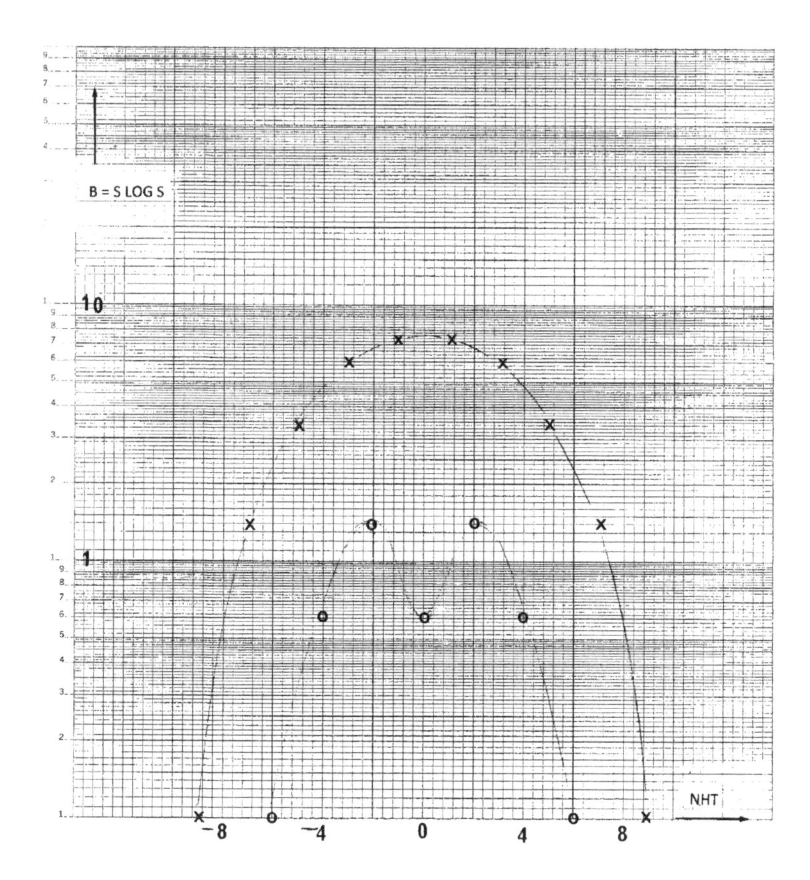

Figure 7-1. Combinatorial information.

second-order fusion, respectively. The dip in the first-order figure at NHT = 0 is due to the need to excise the two disallowable combinations. We will consider the influence of degeneracy due to composition and contingency on information content in the next chapter.

8

Composition and Contingency

To this point we have ignored a major mismatch, namely that between the ***configurational*** approach and the ***direct product/ convolution*** approach to tabulating degeneracy and have emphasized the latter. Now, however, in terms of configuration, suppose we consider the taxonomy from the point of view of the availability of choices imposed by the *basic topology* of each order of fusion. In what follows all "words" (recall the discussion in lexicographic terms in Chap. 7) will be assumed to begin with a basic letter rather than a basic conjugate letter. With that stipulation, the first order fusion of a basic letter and a basic conjugate letter can simply be depicted for our purpose here as in the stick figure of Figure 8-1. We see that the process of fusion has resulted in two quirks, two antiquirks and a junction, five items in total. There is thus the *availability* of 2exp5 = 32 *binary choices*, ***twice*** as many as we found by considering only combinations and permutations.

Second-order fusion can take place in two distinct configurations as suggested by the stick figure representations in Figure 8-2. There are four free quirks, one free antiquark and two junctions in each configuration implying the availability of 2exp7 = 128 binary choices per diagram for a total of 256 choices, ***four times*** the number of terms found on the basis of combination and permutation alone, as in the preceding.

To reconcile the discrepancy, the additional degeneracy due to the factors of ***composition*** and ***contingency*** mentioned above need to be taken into account and combined with the preceding results due to combination and permutation. To begin with, we discuss the number

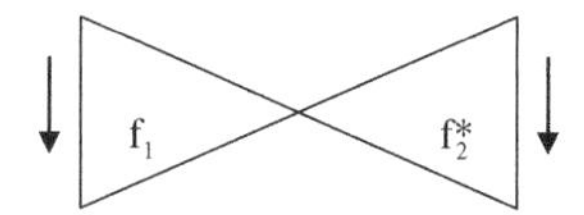

Figure 8-1. Basic operational diagram for first-order fusion.

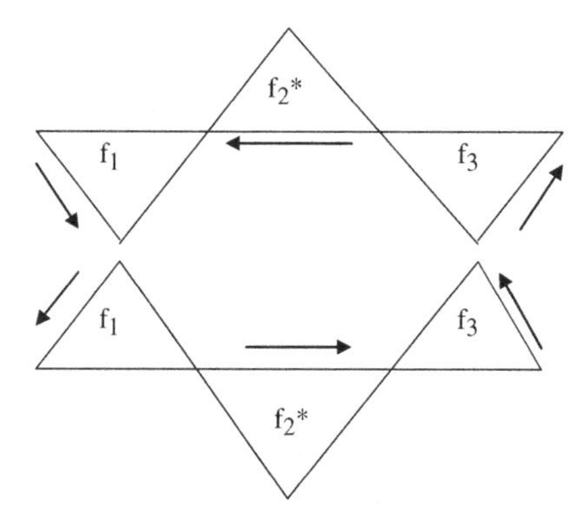

Figure 8-2. Same for second-order fusion.

Table 8-1: Total first-order degeneracy.

AA* (x)	CA* (x)
AB* $(2x)$	CB* $(2x + 2y)$
AC* (x)	CC* $(4y + x)$
AD* (0)	CD* $(2y)$
BA* $(2x)$	DA* (0)
BB* $(4x + y)$	DB* (y)
BC* $(2x + 2y)$	DC* $(2y)$
BD* (y)	DD* (y)

of distinct ways that letters and conjugate letters can combine to form junctions as a result of their *composition* in terms of quirks and antiquirks. In first-order fusion this is relatively straightforward; invoking the basics of Chap. 3 leads to the results summarized in Table 8-1 where the individual terms are listed in dictionary order. In this figure, dd* and uu* junctions are indicated by the letters x and y, respectively in parentheses and the numerical coefficients indicate how many ways the two letters can fuse to form each type of junction.

For example, the three d quirks for letter A (and d* for A*) count as a singlepoint of first-order fusion to form AA* because of the unbroken triangular symmetry of each component. On the other hand, each of the two d* antiquirks of letter B* counts as a potential point of fusion because the symmetry is broken by the direction of traverse and the existence of its u quirk. Hence the term $2x$ for words AB* and BA* which we interpret to mean that those words are each doubly degenerate. In forming BB* the single u quirk of B and the single u* antiquirk of B* can fuse in only one way but the two d quirks of B and two d* antiquirks of B* can fuse in four ways, ergo the term $4x+y$, meaning that BB* is degenerate in five ways. Similar considerations apply in the case of BC* and the two words CB* and CC* in the second column. In that regard, note that inverting, top to bottom, the second column and exchanging x and y as well as A and D, B and C as before makes it identical to the first column. Finally, we note that this is where the exclusionary considerations discussed above come into play. Specifically, we note the "0" terms associated with words AD* and DA* meaning that those words cannot form because neither an x nor a y junction is possible.

Summing up the terms in each coefficient of the words then produces the total degeneracy for first-order fusion as shown in Figure 8-3 where we have regrouped according to twist. Comparing this figure with its original counterpart, Figure 3-4, we see that the consideration of detailed composition has removed two terms (from the NHT = 0 column) but added 18 terms for a net increase of 16

NHT = –6	–4	–2	0	2	4	6
AA*	2×AB*	AC*				
	2×BA*	CA*	4×CB*	5×CC*		
		5×BB*	4×BC*	BD*	2×CD*	
				DB*	2×DC*	DD*
Sum = 1	4	7	8	7	4	1

Figure 8-3. Total degeneracy for first-order fusion.

terms; this doubles the count of Figures 3-3 to 32 which matches the number associated with the configuration of Figure 8-1 and is twice the direct product/convolution tally. Thus we have ***reconciled*** the cited discrepancy for first-order fusion.

We can formally encode this *ad hoc* compilation by invoking the requirement for *continuity of traverse* which implies the mutually exclusive existence of only two kinds of junctions, an x junction formed by the fusion of a d quirk and a d* antiquirk and a y junction formed by the fusion of a u quirk and a u* antiquark, du* and ud* combinations being excluded. On this basis, it is straightforward to formalize the process of fusion as follows: we define $\mathrm{dd}^* = x$, $\mathrm{uu}^* = y$, $\mathrm{du}^* = \mathrm{ud}^* = 0$, the *symbolic* ***two-vectors***

$$\alpha_\mu(\ell \mathrm{d} + m\mathrm{u}) \quad \text{and} \quad (p\mathrm{d}^* + q\mathrm{u}^*)\beta_\nu \tag{8-1}$$

and an ***inner product*** such that

$$\alpha_\mu(\ell \mathrm{d} + m\mathrm{u})(p\mathrm{d}^* + q\mathrm{u}^*)\beta_\nu = \alpha_\mu\beta_\nu(\ell p x + m q y). \tag{8-2}$$

Here, α_μ and β_ν, are α_μ = A, B, C and D and β_ν = A*, B*, C* and D* as defined in connection with Eq. (3-4) of Chap. 3, coefficients ℓ and m are the number of down and up quirks in the FMS, respectively, and p and q are the number of antidown and antiup antiquirks in the conjugate FMS. For example, to compute the structural degeneracy associated with the fusion of a B and a C* we write

$$\mathrm{B}(2\mathrm{d} + \mathrm{u}](\mathrm{d}^* + 2\mathrm{u}^*)\mathrm{C}^* = \mathrm{BC}^*(2x + 2y), \tag{8-3}$$

which we recognize as the relevant entry in Table 8-1.

The effects of composition can be included within the formalism of symbolic convolution with some modification in the definition of the quantities that are convolved. In the case of first-order fusion the modification amounts to changing the α_μ and β_ν letters of Eq. (8-1) from (A, B, C and D) and (A*, B*, C* and D*), respectively, to the enhanced letters (***A***, ***B***, ***C*** and ***D***) and (***A****, ***B****, ***C**** and ***D****), respectively, where each of the first set (of enhanced letters) is the original letter appended by its quirk constituency and each of the second set (of enhanced conjugate letters) is the original conjugate letter preceded by its antiquirk constituency. In other words, we

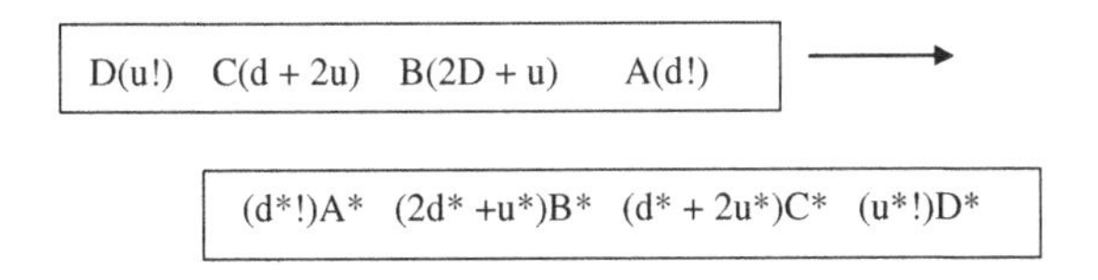

Figure 8-4.

have a pair of two-vectors as defined in the previous section so that, given the restrictions on forming junctions, we simply form the inner product as per Eq. (8-1) to obtain the enhanced product. In fact, Eq. (8-2) is an example of the procedure. That is, the enhanced B is $\boldsymbol{B}$ = B (2d + u) and the enhanced C* is (2d* + u*)$\boldsymbol{C}^*$. The enhanced product is then the inner product BC*$(2x + 2y)$, exactly as shown in Table 8-1.

Figure 8-4 then shows, in the diagrammatic fashion introduced in Chap. 3, the enhanced convolution operation in the case of first-order fusion. The exclamation points are intended to signify that the three d quirks of A count as a single point of fusion in first order fusion as do the three u quirks for D.

We see the operation at a stage poised to generate the permutations

$$\mathrm{A(d!)(d^* + 2u^*)C^*} = \mathrm{AC^*}(x),$$
$$\mathrm{B(2d + u)(2d^* + u^*)B^*} = \mathrm{BB^*}(4x + y) \quad \text{and}$$
$$\mathrm{C(d + 2u)(d^*!)A^*} = \mathrm{CA^*}(x)$$

for the enhanced first stage fusion twist value $n = -2$ which leads to the multiplicity values shown in the -2 column of Figure 8-3, namely AC*, 5BB* and CA*.

The salient feature in ***second order*** fusion is that the junctions *available* for fusion therein depend on junction selection in the *first* fusion, a ***contingency***, which we note in passing, implicates a *Markov* process. The way in which this availability is determined is shown in detail in Appendix C and formalized using the vector algebra developed in the foregoing. Here, we describe the structure and the contingencies of the process as determined simply in an *ad hoc* manner and show how the results are thereby *enhanced.*

Table 8-2: Total second-order degeneracy (but half of all entries).

AA*A $(x)(2x^*)$	BA*A $(2x)(2x^*)$
AA*B $(x)(4x^*)$	BA*B $(2x)(4x^*)$
AA*C $(x)(2x^*)$	BA*C $(2x)(2x^*)$
[AA*D]	[BA*D]
AB*A $(2x)(x^*)$	BB*A $\{(4x)(x^*)+(y)(4x^*)\}$
AB*B $(2x)(2x^*+y^*)$	BB*B $\{(4x)(2x^*+y^*)+(y)(4x^*)\}$
AB*C $(2x)(x^*+2y^*)$	BB*C $\{(4x)(x^*+2y^*)+(y)(2y^*)\}$
AB*D $(2x)(y^*)$	BB*D $(4x)(y^*)$
[AC*A]	BC*A $(2y)(x^*)$
AC*B $(x)(2y^*)$	BC*B $\{(2x)(2y^*)+(2y)(2x^*+y^*)\}$
AC*C $(x)(4y^*)$	BC*C $\{(2x+2y)(x^*+2y^*)\}$
AC*D $(x)(2y^*)$	BC*D $\{(2x)(2y^*)+(2y)(y^*)\}$
[AD*A]	[BD*A]
[AD*B]	BD*B $(y)(2y^*)$
[AD*C]	BD*C $(y)(4y^*)$
[AD*D]	BD*D $(y)(2y^*)$

Table 8-2 lists (with explanation below) ***half*** of all the second-order fusion permutations (including potential exclusions) in dictionary order with coefficients that embody the contingencies — 32 words with eight potential exclusions. With the letter exchanges noted above, a second such figure exists for a total of 64 words with 16 exclusions in accordance with the discussion of direct products above.

As an example of the way in which contingency factors determine the entries of this table, consider the word BB*B in terms of a fusion sequence that proceeds from left to right (although the result is independent of direction): the coefficient in this case $\{(4x)(2x^* + y^*) + (y)(4x^*)\}$ says that the two d quirks in the first B letter and the two d* antiquirks in B* can form an x type junction in four ways in the first fusion. Once this occurs, the remaining d* antiquirk in B* then has the choice of two d quirks in the second B and the u* antiquirk in B* and the u quirk in the second B can form a y junction in one way.

A y junction can also form to begin with (but in one way) between the first B and B* in which case no u* antiquirk is available in B* to form a second y junction. However, a second x junction can still form in four ways as in the previous paragraph. Summing up these alternatives gives a degeneracy of $4 \times (2+1) + 4 = 16$.

Figure 8-5 then presents the (***first half*** of an) associated degeneracy table, including all factors associated with the list of Table 8-2. This time, consideration of detailed structure has removed the eight terms to be excluded from its original counterpart, Figure 5-5 of Sec. II, but added 104 for a net gain of 96, bringing the total for Figure 8-5 to 128, which quadruples the original number. Given the existence of a *second* table with the letter interchanges noted above, the total for second-order fusion is therefore 256 which matches the number associated with the two geometrical configurations of Figure 8-2 and is four times the direct product/convolution tally. Thus ***reconciliation*** is achieved for *second-order* fusion as well. A more detailed discussion of contingency for second-order fusion is carried out in Appendix D.

Keeping the existence of a *second* table with the letter interchanges noted above in mind, we recall the *informational* content associated with the *combinatorics only* as displayed in Figure 6-1 and construct a similar display associated with the *totality* of the effects of combinatorics, composition and contingency, Figure 8-6 that includes *all the information* contained in Figure 7-5 above plus its dual (not shown). Clearly, there is a lot of information contributed by the extra factors of composition and contingency.

NHT =	–9	–7	–5	–3	–1
	2AA*A	4AA*B 2AB*A 4BA*A	6AB*B 8BA*B 6BB*A 2AA*C 2CA*A	6AB*C 4CA*B 2BC*A 6CB*A 2AC*B 4BA*C 16BB*B	4AC*C 2CA*C 4CC*A 2AB*D 2DB*A 14BB*C 10BC*B 14CBB*
SUM =	2	10	24	40	52

Figure 8-5. Total twist-charge grouping (all degeneracy but half of entries).

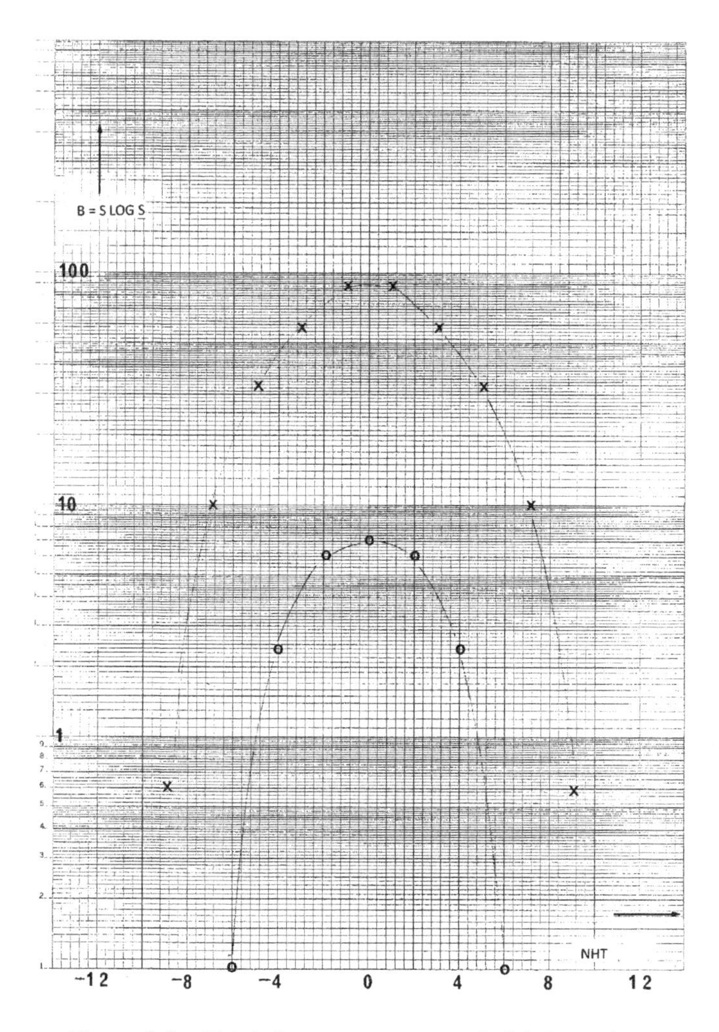

Figure 8-6. Total degenerate taxonomical information.

And, finally, one last bit of perspective, something that applies to all the preceding, governing all the taxonomical development we talked about; it's the simple but fundamental requirement that an up quirk can mate only with an up antiquirk and similarly a down quirk only with a down antiquirk (u with u* and d with d*). Does this remind you of anything? How about DNA? You may recall that a similar, ***self-complementary*** feature — Guanine mating only with

Cytosine and Thymine only with adenine (G with C and T with A) is what allows DNA to implement the reproduction process making it conceivable that, some day, I would be here to write this book and you would be here to read it! And how about that "fourfoldiness" in each case right up front!

I found all that intriguing, even thought provoking! So, it did provoke some thoughts and I put them down in Appendix F. "The Alternative Model and Dark Matter Redux and DNA", turned the crank and what came out was a principle, one that appears to be of vast generality. I call it the Principle of Complementarity and talk about it at some length in the concluding remarks of Sec. VII.

III
Algebraic Ruminations

Although this section is not absolutely crucial to the main subject matter of the book it provides some useful additional perspective. Some of it gets a little messy so if you are so minded you can skip it but I would recommend that you at least see what is involved. I trust you will be glad you did but there are no refunds if not satisfied. There are six subjects, all algebraic to one degree or another and they cover a large range of topics, all germane to the book's basic message.

9

The Bialgebra and Hopf Algebra Connection

Here we develop a relationship between the preceding material, in particular the matter of fusion and fission, and what's known as a ***Bialgebra*** and a ***Hopf*** *algebra* [26, 30], the latter often discussed under the subject of "*Quantum Groups*". The notions of fusion and fission being clearly central to FMS modeling, the fact that they take place explicitly between particular quirks and their antiquirk counterparts bears reemphasizing. The subject was treated in Sec. II in terms of first, what was called symbolic convolution, and then a vector formalism, both unique to this book and its predecessors [10–14].

However, it is gratifying that there exists a well-known (*tensor*) formalism, the Hopf Algebra, that also applies to the geometrical manipulations involved therein. Figure 9-1 below (repeated from Figure 2-10 of Sec. II) shows diagrammatically the allowable, quirk–antiquirk pairs, both in the ***free*** and ***fused*** conditions, first in plan view and then in the corresponding edge views. The edge views portray in a schematic way the connections being made ***vertically*** (relative to the page) in the free condition and ***horizontally*** in the fused condition.

By way of comparison, Figure 9-2 summarizes, in a top level, schematic way, the Hopf algebra's operations of ***multiplication*** and ***comultiplication*** [e.g. 28].

In terms of a diagrammatical summary, these two figures basically say the same thing: on the one hand ***two*** entities combine to make a ***third***, and on the other ***one*** entity splits into ***two***. Briefly, if we can associate the process of ***fusion*** with the operation of ***multiplication***

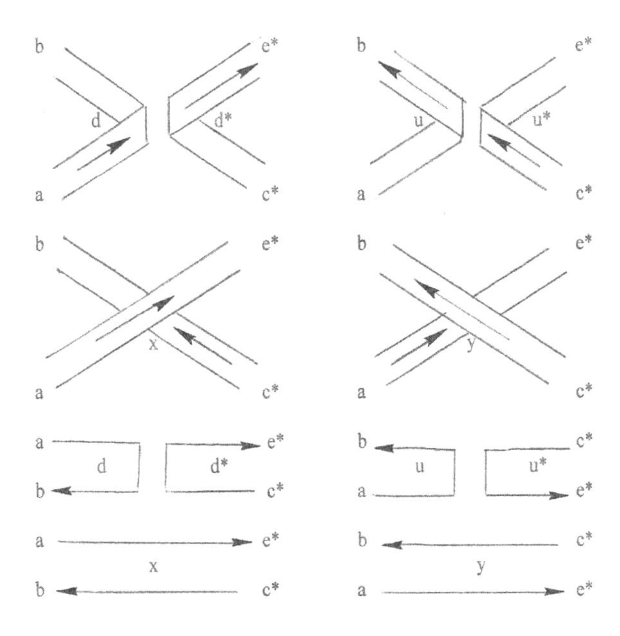

Figure 9-1. Free and fused quirk-antiquirk pairs.

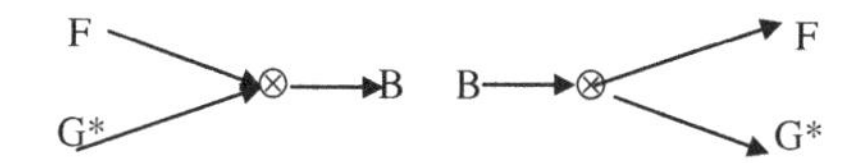

Figure 9-2. Hopf multiplication and comultiplication.

and the process of ***fission*** with the operation of ***comultiplication*** we will have endowed our alternative model with two essential components of a ***bialgebra***, and with the inclusion of an ***antipode***, the essential elements of a Hopf algebra [26, 28, 30]. Although the word "antipode" evokes geometrical associations — for example, the north and south poles of the Earth are in an antipodal relationship — the Hopf concept is germane strictly to algebraic systems wherein it is often described as a kind of ***inverse***. It is not a simple concept and understanding its connection herein really requires considerable background development. For that purpose the treatment in [26] was relied upon but for the purpose of this book can only be abstracted (see below), the bulk of the discussion in what follows being devoted to the bialgebra connection. Nevertheless, it turns out that the antipode associated with the algebra of taxonomical development as carried out in the preceding is quite gratifyingly straightforward as we shall see.

As indicated above, what's important in a *Hopf algebra* is that the operations of *multiplication* and *comultiplication* are representative, of corresponding ***tensor*** operations [30]. From our model's point of view, the inputs F and G^* are *basic fermions* and *antifermions,* respectively and the Bs are ***bosons***, members of matrix M (Sec. II, Chap. 3). One requirement for relating our model to a Hopf algebra is therefore to model the fermion and antifermion fusion and fission operations in a *tensor* formalism. Actually, this representation began in [10] and was more fully developed in [11] well before the Hopf connection discussed herein was initiated but in any event, the idea is to represent the basic ***fermion*** FMS as ***covariant*** tensors and the corresponding ***antifermion*** FMS as ***contravariant*** tensors. I apologize for what may become a bewildering array of subscripts, superscripts and definitions; that seems to be the nature of the beast.

To begin with, we express the close coexistence (not yet fusion) of a free, basic fermion and antifermion FMS pair as

$$\mathrm{C}_{abc}^{r^*s^*t^*} = F_{abc}G^{r^*s^*t^*}, \tag{9-1}$$

where F can be either A, B, C or D; G can be either A*, B*, C* or D*; each of a, b and c can be either d or u and each of r, s, and t can be either d* or u*. Implementing ***fusion*** is then expressed by equating a ***subscript*** of F and a ***superscript*** of G using the ***Kronecker delta*** function and the summation implied in subscript–superscript identification as

$$\delta_n^m F_{abc}G^{r^*s^*t^*} = B_{abc/m}^{r^*s^*t^*/n}, \tag{9-2}$$

where m can be either a, b or c and n can be either r^*, s^* or t^*. The subscripts abc/m and superscripts $r^*s^*t^*/n$ on the R.H.S are meant to imply that, upon the choice of values for m and n, two subscript and two superscripts remain for B. Thus, B becomes the representation of a two-quirk, two-antiquirk boson. For example, with

$m = a$ and $n = r^*$ we have

$$\delta_r^a F_{abc} G^{r^* s^* t^*} = \mathrm{B}_{bc}^{s^* t^*}. \tag{9-3}$$

Note, however that, if one were to view the R.H.S. by itself, something would seem to be missing, namely, the nature of the crossover junction, that is, whether it's an x (for dd*) or a y (for uu*) as well as knowledge of which FMS were involved in the fusion to begin with. To rectify that deficiency, we define the fusion ***state function***, $\langle \Phi_{v(m)} \rangle$ (see below) treating it as an ***operator*** and evaluating it as per the delta function — that is, instead of Eq. (9-2) we would write

$$\begin{aligned} \delta_n^m \langle \Phi_{v(m)} \rangle F_{abc} G^{r^* s^* t^*} &= F_{abc/m} G^{r^* s^* t^* /n} (z_{v(m)}) \\ &= B_{bc/m}^{s^* t^* /n} (z_{v(m)}), \end{aligned} \tag{9-4}$$

where $v(m)$ means the ***value*** of m which can be either d or u, and further, z can represent either an (x) or a (y) junction *depending upon* whether v is a (d) or a (u), respectively. For example, with $m = a$, $n = r$ as in the above and, further, with $a = d$ and, correspondingly, $r = d^*$, (an x junction) Eq. (9-4) evaluates as

$$\delta_r^a \langle \Phi_{v(a)} \rangle F_{abc} G^{r^* s^* t^*} = \mathrm{B}_{bc}^{s^* t^*} (x). \tag{9-5}$$

In essence, the delta function indicates which quirk–antiquirk pair is to be fused (in this case it is the d/d* pair) and the state function does the fusing, leaving its mark, the value of the junction, in the process.

In the general case, we would write the R.H.S. of Eq. (9-5) as $\mathrm{B}_{bc}^{s^* t^*}(m_z)$, where $m = d$ or u depending upon whether z is x or y, respectively. So, now, if we encounter a boson represented in that form we would know everything about it, that is to say the complete makeup of each of its constituents, and we should like that representation to be susceptible to a representation of the fission process. For that purpose we invoke the mate to $\langle \Phi_\nu \rangle$, the fusion operator, namely the fission state function, $\langle \phi_\alpha \rangle$, combine it with the delta function, δ_z^α, and apply it to the generalized R.H.S. of Eq. (9-5), $\mathrm{B}_{bc}^{s^* t^*}(m_z)$. The situation here, is that, since we ***know*** the values of both z and m, what we need to do is to make sure that the fission operator knows it has to separate a z-type junction. That is, we want $\langle\phi_\alpha\rangle$ to become $\langle\phi_z\rangle$ so we write

$$\mathrm{B}_{bc}^{s^*t^*}(m_z)(\delta_z^\alpha\langle\phi_\alpha\rangle) = \mathrm{B}_{bcm}^{s^*t^*m^*}, \tag{9-6}$$

where the R.H.S. is now in the form of the coexistence of a ***free***, basic fermion and antifermion pair as in Eq. (9-1). That is, we can write

$$\mathrm{B}_{bcm}^{s^*t^*m^*} = F_{bcm}G^{s^*t^*m^*} \tag{9-7}$$

as in Eq. (9-1).

Regarding the ***state functions*** introduced above, we recall the edge views of Figure 9-1 above that show quirk connections in fusion being made "vertically" (relative to the Page) in the free condition and "horizontally" in the fused condition. This is the fundamental dichotomy — the connections in this representation are either horizontal or vertical. Thus for fusion to occur, the two vertical connections must be disenabled and the horizontal connections must be enabled and, conversely, for fission to occur, the two horizontal connections must be disenabled and the two vertical connections must be enabled. To pursue that line of thought, we define ***state functions*** for the ***horizontal*** and ***vertical*** connections of a quirk–antiquirk pair to be $\langle\Psi_\alpha\rangle$ and $\langle\psi_\alpha\rangle$, respectively, where the value of α is either d or u. Also, we define A and $A^\dagger$ to be the ***disenabling*** and ***enabling*** operators, respectively. Then, the *fusing* of a junction between an originally-free FMS and antiFMS pair can be expressed in terms of a state function as

$$\langle\Phi_\alpha\rangle = \mathrm{A}\langle\psi_\alpha\rangle + \mathrm{A}^\dagger\langle\Psi_\alpha\rangle. \tag{9-8}$$

Also, we define the ***value*** of $\langle\Phi_\alpha\rangle$ to be (z_α) such that $z_d = x$ and $z_u = y$, respectively. Similarly, the freeing-up of an originally-fused pair can be expressed in terms of a similar state function as

$$\langle\varphi_\alpha\rangle = \mathrm{A}\langle\Psi_\alpha\rangle + \mathrm{A}^\dagger\langle\psi_\alpha\rangle. \tag{9-9}$$

Although the analogy of the disenabling and enabling operators above to the ***annihilation*** and ***creation*** operators of quantum field theory (QFT) was perpetrated with malice aforethought, we will not pursue the subject at this time. However, later on in the book we shall discuss the relationship of our model to topological quantum field theory (TQFT) (which is not exactly the same thing).

Regarding the ***antipode***, as per the above, we begin by abstracting the treatment in [25] which was developed in connection with the so-called "***quantum group***", $\mathrm{SL}(2)_q$ as follows: first a matrix operator

$$P = \begin{pmatrix} a & b \\ c & d \end{pmatrix}, \tag{9-10}$$

is defined such that $P\tilde{\varepsilon}P^{\mathrm{T}} = \tilde{\varepsilon}$ and $P^{\mathrm{T}}\tilde{\varepsilon}P = \tilde{\varepsilon}$ where $\tilde{\varepsilon} = \begin{pmatrix} 0 & \mathrm{A} \\ -\mathrm{A}^{-1} & 0 \end{pmatrix}$ is the ***defining invariant*** for the "quantum group" and A, which appears, as below, in the well-known ***bracket solution*** [9, p. 29] to the ***Yang–Baxter*** equation,[1] namely, in Knot theoretic terms,

$$\text{(crossing } a, b, c, d) = \mathrm{A}\,\big|\,\big| + \mathrm{A}^{-1} = , \tag{9-11}$$

is assumed to ***commute*** with the elements of P. Next, as a result of conditions imposed on matrix P, it is found that the ***connected entries***, and ***only*** those in the diagram on the L.H.S. of the Yang–Baxter equation commute (that is a with d and b with c), that is

$$\begin{aligned} da - ad &= 0, \\ bc - cb &= 0. \end{aligned} \tag{9-12}$$

Finally, the associated ***antipode*** is ***defined*** [18, p. 142] so that the matrix ***inverse to*** P is

$$P^{-1} = \begin{pmatrix} d & b \\ c & a \end{pmatrix}, \tag{9-13}$$

i.e., the ***positions*** of entries a and d are ***interchanged***, in essence the "bottom line" of the development.

---

[1] An important approach to solving statistical mechanics problems, for instance two-dimensional arrays of spins that are either "up" or "down", or the Kauffman analog of crossings encountered in traversing a knot displayed in the usual two-dimensional array [26].

For our model, the approach here is therefore to associate matrix P with a ***particular*** FMS and its matrix ***inverse*** with the FMS it consequently implies. Comparing the two FMSs should then allow the nature of the antipode to emerge. For the reference FMS we choose the configuration indicated in the top right of Figure 9-1 (although either of the two will do) which we ***abstract*** but annotate as shown, recognizing its nature as a bound state of a fermion and an antifermion.

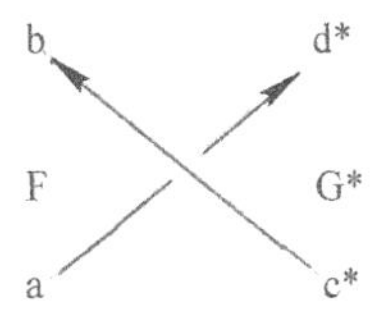

Rotating this diagram 90° cw produces the geography of matrix P annotated with directional indications.

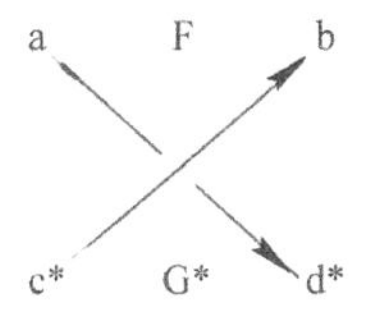

whose complex conjugate (the view from the rear as per the Wheeler–Feynman notion; See Sec. VI, Chap. 30) is the diagram

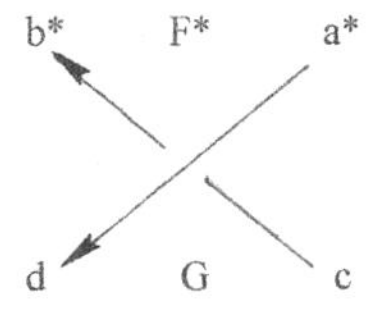

which, rotated 90° cw is indeed the annotated version of inverse matrix P^{-1} as we see below. And, furthermore, it's conjugate converse (as seen in a rear view) is the reference schematic we started with.

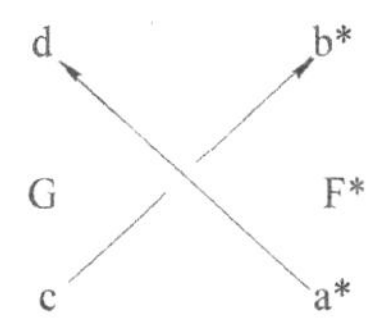

In other words, the ***inverse*** matrix in the Hopf formalism implicates an FMS ***conjugate*** to our arbitrarily chosen schematic of a first-order fusion FMS and we conclude that the ***antipode*** to a first-order fusion bound state of our model is simply its ***conjugate***! Actually, those pairs of elements of matrix M that are antisymmetrically deployed about the twist gradient are ***each*** in an ***antipodal*** relationship, for example, conjugate pairs (AB* and BA*) which (as we will see in the next chapter) correspond to the vector boson pair (W^- and W^+) and the pair (BC* and CB*) which correspond to the charged pions we saw above.

10

A Foray into Quantum Mechanics

This chapter considers the possibility of giving the model a quantum mechanical treatment. Since this book is not concerned with particle kinematics, position/momentum relationships are not an issue. However, there is clearly a statistical element in the way FMS are combined in fusion and fission operations and this should rightly be treated in a quantum mechanical manner. Of interest in that regard is the remark of Sec. II, Chap. 3 that matrix M can be viewed as an ***operator*** that converts a vector representing the set of four basic, single component (letter) fermions into a vector of four matrices, one that codifies the set of three letter words representing the set of three-component fermions. In this sense, it invokes the ***quantum mechanical*** notion that the ***outer product***, say $|\beta\rangle\langle\alpha|$ of a *ket*, $|\beta\rangle$, and a *bra*, $\langle\alpha|$, is an *operator* that converts a *ket* into another *ket*.

Here, of course the roles of bras and kets are played by our fermion and antifermion FMS and the detailed operator role is actually *implemented* by the individual elements, that is to say the ***bosons*** of matrix M which are indeed constructed as outer products. On a more fundamental level, we note that a salient feature of quantum mechanics is its formulation in terms of ***complex algebra***. Coincidentally, a salient feature of matrix M, in addition to its role as an operator, is the ***orthogonality*** of its ***twist*** and ***charge*** gradients mentioned originally in Chap. 3 of Sec. II, which we illustrate in Figure 10-1 in terms of the gradients prevailing in that manner in second order fusion as well.

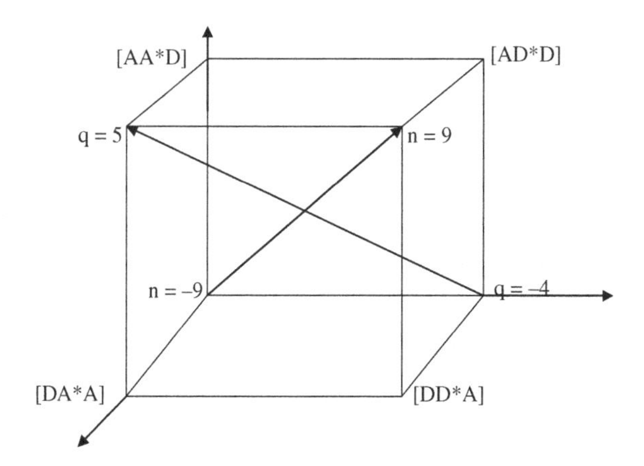

Figure 10-1. The orthogonal twist-charge gradients for second-order fusion.

To take advantage of this basic orthogonality, it appears reasonable to consider a ***complex reformulation*** involving both twist and charge. One way to do this is in terms of an FMS ***state function*** (not the fission/fusion state functions as in the previous chapter!). We begin by equating each *basic* ***fermion***, call it, generically x, with a ***state***, one that features a corresponding complex ***twist/charge state vector*** that explicitly features twist-charge orthogonality.

$$C(x) = t(x) + jq(x) \tag{10-1}$$

and each basic antifermion, say y^*, with a corresponding conjugate state vector.

$$C(y^*) = t(y^*) - jq(y^*). \tag{10-2}$$

(Note the correspondence to the usual canonical position-momentum variables.)

In formulating the fusion process, we would like to combine states in a ***multiplicative*** fashion in line with the *lexicographic* way the fusion process has heretofore been characterized, while at the same time allowing twist and charge information to be handled ***additively*** thus mirroring actual FMS combinatorics. Accommodating both requirements suggests an ***exponential*** formalism. That is, we would formulate something like

$$xy^* = e^{C(x)}e^{C(y^*)} \tag{10-3}$$

such that what appears in each exponent is in the nature of the ***phase*** of a vector in a ***complex*** *domain.* The matter of an associated *amplitude* will be discussed below, specifically, as we shall see, in terms of the *combination of* degenerate states.

From a slightly different point of view, we note that each of these exponential functions is actually the ***Fourier*** *Transform* of a ***delta*** *function* located at the terminus of an associated complex twist/charge vector in the n/q domain as illustrated in Figure 10-2 which represents the three charge degeneracies for $n = 2$ and the two charge degeneracies for $n = 4$ that we found in the second order fusion matrix M. Clearly, fixing *twist* fixes the ***real*** component of each of the associated complex constituent vectors, which therefore differ only in their ***imaginary*** (*charge*) components. The locus of the end points of those vectors for a given value of twist is therefore a vertical line. Thus a given location and its transform are just *complementary* ways of characterizing the nature of a given FMS.

The transform can also be used to illustrate another connection to a quantum mechanical formalism, namely the algebra of ***commutativity*** which, we note, is related to the Poisson bracket formalism that provides a bridge between classic and quantum

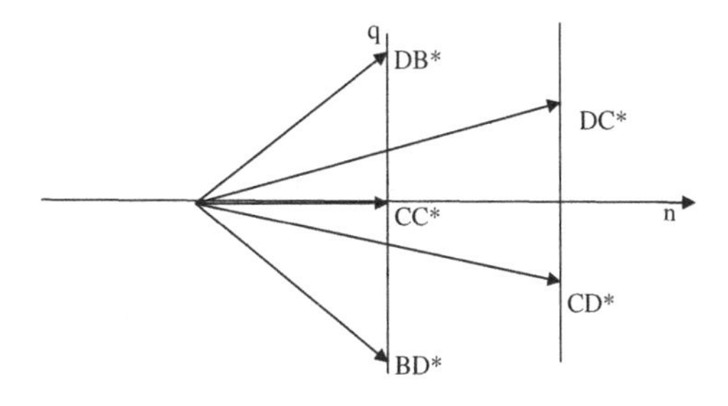

Figure 10-2. Example of complex second-order twist-charge vectors.

analysis. Consider the commutator (actually expressing the ***fusion*** of x and y^* or y and x^* to create ***bosons***).

$$\begin{aligned}
\mathbb{C}om &= e^{C(x)}e^{C(y^*)} - e^{C(y)}e^{C(x^*)} \\
&= e^{(n_1+jq_1)}e^{(n_2-jq_2)} - e^{(n_2+jq_2)}e^{(n_1-jq_1)} \\
&= e^{(n_1+n_2)}[e^{j(q_1-q_2)} - e^{-j(q_1-q_2)}] \\
&= 2je^n \sin(q_1 - q_2) \\
&= 2je^n \sin(n_1 - n_2),
\end{aligned} \tag{10-4}$$

where, as before, $n = n_1 + n_2$. Clearly, we have commutation ($\mathbb{C}om = 0$) only for $x = y$.

We also need to address the formulation of state functions that incorporate the *degenerative* states associated with a given value of twist. Although fixing twist means we can group degeneracies by charge, we must still account numerically for the multiplicities due to the several *sources* of degeneracy, as per the discussion in Sec. II. Charge grouping for first order fusion is simple (we have already seen part of it); only *individual* charge values for each value of twist exist — no groups. However, the *detailed* composition of these constituents in terms of quirks and antiquirks allows some of them to be replicated in more than one way. The net result was that what we might term a ***replication*** *factor* was attached in a multiplicative way to ***augment*** each of the listed constituents of first order fusion as was shown in Figure 8-1 of Sec. II. Taking replication due to charge *plus* detailed composition into account then suggests a state function in the following form:

$$\begin{aligned}
XY^* &= \sum_{i=1}^{Q(n)} \rho_i e^{(n+jq_i)} \\
&= e^n \sum_{i=1}^{Q(n)} \rho_i e^{jq_i},
\end{aligned} \tag{10-5}$$

where

$$\begin{aligned} n &= \text{the value of twist} \\ Q(n) &= \text{the number of associated charge constituents} \\ q_i &= \text{the value of the } i\text{th charge} \\ \rho_i &= \text{the number of distinct compositional} \\ &\quad\ \text{replications for the } i\text{th charge.} \end{aligned}$$

For example, for $n = 2$, we have three combinatorially degenerate states, each with its own charge and number of distinct replications, namely (see Figure 10-2)

$$\begin{aligned} &\mathrm{CC}^*(q = 0, \rho = 5) \\ &\mathrm{BD}^*(q = -2, \rho = 1) \\ &\mathrm{DB}^*(q = 2, \rho = 1) \end{aligned}$$

so that Eq. (10-5) evaluates as

$$\begin{aligned} XY^* &= e^2(5e^{j0} + e^{-j2} + e^{j2}) \\ &= e^2(5 + e^{-j2} + e^{j2}). \end{aligned} \tag{10-6}$$

Second order fusion is understandably somewhat more complicated: we write

$$XY^*Z = e^n \sum_{i=1}^{Q(n)} e^{jq_i} \sum_{k=1}^{R(n,i)} \rho_{ik}, \tag{10-7}$$

where

$$\begin{aligned} n &= \text{value of total twist} \\ Q(n) &= \text{number of associated charge } \textit{groups} \\ R(n,i) &= \text{number of distinct combinatorial members of the} \\ &\quad\ i\text{th } \textit{group} \\ \rho_{ik} &= \text{number of compositional } \textit{replications} \text{ of the} \\ &\quad\ k\text{th member of the } i\text{th charge group.} \end{aligned}$$

For example, for the case of $n = -3$ (see Figures 5-3 and 8-5) there are three charge groups

$$\begin{aligned}
q = -2 \text{ with two members,} \quad & \text{AC}^*\text{B with } \rho = 2 \\
& \text{BC}^*\text{A with } \rho = 2 \\
q = 0 \text{ with three members,} \quad & \text{AB}^*\text{C with } \rho = 6 \\
& \text{BB}^*\text{B with } \rho = 16 \\
& \text{CB}^*\text{A with } \rho = 6 \\
q = 2 \text{ with two members,} \quad & \text{BA}^*\text{C with } \rho = 4 \\
& \text{CA}^*\text{B with } \rho = 4
\end{aligned}$$

so that Eq. (10-7) evaluates as

$$\begin{aligned}
XY^*Z &= e^{-3}\left\{e^{-j2}(2+2) + e^{-j0}(6+16+6) + e^{j2}(4+4)\right\} \\
&= e^{-3}(28 + 4e^{-j2} + 8e^{j2}). \qquad (10\text{-}8)
\end{aligned}$$

In summary, the foregoing demonstrates a formulation that combines complex algebraic precepts of quantum mechanics with the degeneracies and combinatorics inherent to the FMS genus. There is a connection to what is known as the Wigner model of quantum mechanics but all such further development awaits the next book.

11

Spin, Spinors, and the Pauli Connection

We begin this chapter with a quotation: "No one fully understands spinors. Their algebra is formally understood but their general significance is mysterious. In some sense they describe the "square root" of geometry and, just as understanding the square root of -1 took centuries, the same might be true of spinors." The speaker, who is a mathematician of the highest repute in many areas including the field of particle physics and its relationships shall be nameless, mainly because I seem to have mislaid the reference for this rather gloomy prognostication. Were he to become aware of it, I should hope he would forgive me for saying so, but I think we may be able to do better than that; I don't think I can afford to wait centuries and perhaps the rest of this chapter will indicate why we do not really need to.

I recognize, of course, that the general subject of spin and spinors is indeed complex and that numerous types of spinors with some celebrated names attached to each have been defined to explicate various physical and mathematical situations. However, that's not our concern in this book. Our concern is mainly with the elementary particles of our alternate model and, in that regard, the characterization of those particles given in the preceding bears repeating: "they are to be regarded not as discrete, point-like objects in a vacuum, nor as quanta of a field but, rather, as *localizably-sustainable* ***distortions*** in and of an otherwise featureless continuum. Indeed, they are not seen as "***objects***" at all in the ordinary sense of the word, but as *topological entities* — ***solitons*** — which persist because they are *topologically nontrivial*; they cannot dwindle away to a point and disappear." From this point of view, alternative model particles are ***classical*** (*not* ***quantum***) entities. Nevertheless, a most important, intrinsic consequence of the modeling (see below) is that they ***manifest*** the fundamental ***quantum*** *attribute of* ***spin***. In fact, they constitute manifestations of ***spinors***. And as remarked in the preface to a textbook on the subject [32], "— the foundations of the concepts of spinors are groups; spinors appear as representations of groups." Which implies that ***our particles*** also constitute manifestations of groups, or more explicitly, it turns out, a ***particular group***, namely the gauge group SU(2).

This is an interesting situation: in the quantum mechanical model of the elementary particles, the concept of the attribute of spin was originally postulated in 1925 by Goudsmit and Uhlenbeck [33], in order to explain the fine structure observed in atomic spectra in the presence of a magnetic field and was then formally elucidated in terms of the Pauli/Dirac theory. A basic element of this concept is the theory of ***spinors***, entities which were actually introduced by **Cartan** [31] in 1913, well before they made their way into particle physics. In our model, however, as per the above, they emerge as ***classically-describable*** entities inherent to the ***toroidal ontology*** of the MS itself.

SU(2), the group of special, unitary, 2×2 complex matrices, is well known to map onto the group SO(3) of rotations in 3-space in a two-to-one manner, that is, such that a rotation by angle θ, expressed as an SU(2) matrix operation on the 2×2, self-adjoint matrix

$$X = x_k\sigma^k = \begin{bmatrix} z & x - iy \\ x + iy & -z \end{bmatrix}, \qquad (11\text{-}1)$$

was noted by Cartan to be ***equivalent*** to a net rotation by 2θ of the ***vector*** $\mathbf{x} = (x, y, z)^T$ in 3-space. Here the σ^k are the ubiquitous **Pauli** (should really be Cartan/Pauli) spin matrices (see below) which, together, constitute a three-component vector whose inner product with the x^k is Eq. (11-1). This circumstance is customarily used in explicating the unusual nature of spin which, although it is measured in units of angular momentum, occurs ***quantized*** *in all* ***half-integer*** rather than only in ***integer*** multiples of $h/2\pi$ where h

is Planck's constant. We saw a number of informal (mainly graphical) manifestations of this very circumstance in Sec. II in terms of the basic concepts and development of our particle model.

Interjection: We note that the Pauli matrices,

$$\sigma^1 = \begin{pmatrix} 0 & 1 \\ 1 & 0 \end{pmatrix}, \quad \sigma^2 = \begin{pmatrix} 0 & i \\ -i & 0 \end{pmatrix}, \quad \sigma^3 = \begin{pmatrix} 1 & 0 \\ 0 & -1 \end{pmatrix}, \tag{11-2}$$

occupy a pivotal position in the development of a variety of mathematical formalisms central to physical theories. As per the characterization above they are indeed ubiquitous. Not only that but we see them all over the place. One might go so far as to characterize them as ***central to the entire subject of spin, spinors and the associated group SU(2).***

For one thing, it is readily demonstrated [35] that they constitute a basis for the Lie algebra associated with SU(2). However, there is more to their claim to fame, some of which we discuss below.

Also, we recall, the development of our ***taxonomy*** was systematized in Sec. II, in terms of the group SU(2) as per the development in [15]. The reason for the choice of SU(2) was stated therein as being because (following Wigner) [15], it is the so-called "little group" associated with the orbits of particles whose kinematics, relativistically speaking, are located in the (relativistically accessible) "forward light cone", with energy–momentum vectors

$$p = (p_0, p_1, p_2, p_3)^T$$

such that $p_0^2 > 0$ and

$$m^2 = p_0^2 - (p_1^2 + p_2^2 + p_3^2) > 0,$$

where m is particle mass, and (2) all its irreducible representations are parametrized by spin in multiples of 1/2. Here, the matrix-equivalent to the Lorentz 4-vector, is formed by substituting $p_1, p_2,$ and p_3 for $x, y,$ and z, respectively, in Eq. (11-1), and adding p_0 to z.

All of which is mainly of academic interest only (by way of justification for the group theoretic summary of our taxonomical development in Sec. II) because we are done with taxonomy for now and, as stipulated in the previous chapter, we will not be considering kinematics. However Eq. (11-1) ***is*** of direct interest to our model because, in the first place, as per Cartan [33], it constitutes a ***spin matrix***, whose elements define the fundamental two component ***spinor***

$$\xi_0 = \sqrt{(x - iy)/2},$$
$$\xi_1 = \sqrt{(x + iy)/2}. \tag{11-3}$$

Note that the diagonal (that is, the z) elements of the spin matrix are not involved in this formalism, which implies that spin is an essentially ***planar*** phenomenon. At the same time we recall that our concern here is with the notion of the MS as a concatenation of torus ***knots***, each of which, for visualization purposes, can be thought of as a real string wound around a real torus. With reference to Figure 2-6 of Sec. II, the Cartesian coordinates of a point on such a string are (and here we are looking ahead a bit to the differential geometry of Sec. V):

$$\begin{aligned} x &= w\cos\phi, \\ y &= w\sin\phi, \\ z &= r\sin\theta, \end{aligned} \tag{11-4}$$

where $w = R + r\cos\theta$ is the projection in the x, y plane of the radius vector from the origin of coordinates to the point (ϕ, θ) on the toroidal surface, R is the radius of the toroid's circular centerline, r is the radius of its circular cross section and ϕ and θ are angular measures in the azimuthal and meridianal directions, respectively. Then we can translate the spin matrix of Eq. (11-1) as

$$X = \begin{bmatrix} r\sin\theta & w(\cos\phi - i\sin\phi) \\ w(\cos\phi + i\sin\phi) & -r\sin\theta \end{bmatrix} \tag{11-5}$$

and the associated ***spinor*** components as

$$\xi_0 = \sqrt{w(\cos\phi - i\sin\phi)/2} = \sqrt{(w/2)e^{-i\phi}} = e^{-i\phi/2}\sqrt{(w/2)},$$

$$\xi_0 = \sqrt{w(cos\phi + i\sin\phi)/2} = \sqrt{(w/2)e^{+i\phi}} = e^{+i\phi/2}\sqrt{(w/2)}. \tag{11-6}$$

Thus, noting that spin has to do only with circulation in the xy plane, we see that the half-angle exponent on the R.H.S. of Eq. (11-6) implies that a nominal traverse of 2π around the MS results in an increment of only π in the phase of the spinor components. In other words, the MS provides a visualizable ***manifestation*** of the notion of a spinor (which, of course, we had already discussed in Sec. II but not quite in terms of this formalism).

Note, however, that the correspondence is not quite complete being complicated by the presence of the term involving $w = R + r\cos\theta$. Basically, this introduces what may be described as a "modulation" of the projection of the MS in the x, y plane as a function of ϕ. To see this a little more explicitly, note that the equations for the position vector components written out in detail are

$$\begin{aligned} x &= R\cos\phi + \cos\theta\cos\phi, \\ y &= R\sin\phi + \cos\theta\sin\phi, \\ z &= r\sin\theta. \end{aligned} \tag{11-7}$$

Clearly, x, y and z are all periodic functions of ϕ. However, while the variation of z is just a single sinusoid, we can view how x and y vary in the sense of ***communication*** system theory [35] as the amplitude modulation of a carrier signal $R\cos\phi$ by the modulating function $(1 + \rho\cos\phi)$ where $\rho = r/R$. That is, we would write

$$x = R(1 + \rho\cos\phi)\cos\phi \tag{11-8}$$

and similarly for y with $\sin\phi$ instead of $\cos\phi$. To make the correspondence to a spectrum of frequencies, we can write

$$\phi(\ell) = 2\pi m\ell/L, \tag{11-9}$$

where ℓ is the index of ϕ values, $1 \leq \ell \leq L$, such that $\phi(L) = 2\pi m$. Consequently, the equation for x can be rewritten as

$$x = R(1 + \rho \cos 2\pi f_n \ell) \cos 2\pi f_m \ell, \tag{11-10}$$

where $f_m = m/L$ and $f_n = n/L$.

The spectrum implied here is well-known to consist of a carrier signal at frequency, f_m, and a pair of sidebands at $f_+ = f_m + f_n$ and $f_- = f_m - f_n$, a fact readily demonstrated by rewriting the above expression in terms of a trignometric identity as

$$x = R \cos 2\pi f_m \ell + r/2(\cos f_+ \ell + \cos f_- \ell). \tag{11-11}$$

Note that the sideband contribution continues to dwindle down as r/R goes to zero. However, the basic topological nature of the MS, we might say its ***character***, does ***not***. In other words, the toroidal winding of the MS boundary is ***essential*** to the separate ***identification*** of the individual MS regardless of the ratio r/R.

New subject: You may recall the "interjection" above where we mentioned the widespread use of the Pauli spin matrices. A most important instance of that is in the development of the **Dirac** theory of the electron, where we saw in Sec. II how the ability of the Dirac equation,

$$-DD^* \psi = (\gamma^\mu \gamma^\nu \partial_\mu \partial_\nu + m^2)\, \psi = 0, \tag{11-12}$$

to constitute a manifestation of the Klein–Gordon equation depended on the 4×4 gamma matrices satisfying the requirements of a Clifford algebra [17], namely that

$$(\gamma^0)^2 = I^4, \ (\gamma^i)^2 = -I, \quad \text{for } i = 1, 2, 3, \tag{11-13}$$

and

$$\gamma^\mu \gamma^\nu + \gamma^\nu \gamma^\mu = 0$$

where I^4 is the unit 4×4 matrix. A standard realization of this requirement is

$$\gamma^0 = (\sigma^0 \otimes I) = \begin{pmatrix} I & 0 \\ 0 & I \end{pmatrix} \quad \text{and} \quad \gamma^\mu = i(\sigma^2 \otimes \sigma^\mu) = i \begin{pmatrix} 0 & \sigma^\mu \\ -\sigma^\mu & 0 \end{pmatrix} \tag{11-14}$$

for $\mu = 1, 2, 3$, and the sigmas are the Pauli matrices. But, since it is readily verified (by explicit multiplication) that

$$\sigma^{\mu}\sigma^{\nu} + \sigma^{\nu}\sigma^{\mu} = 0, \tag{11-15}$$

it follows that

$$\gamma^{\mu}\gamma^{\nu} + \gamma^{\nu}\gamma^{\mu} = 0 \tag{11-16}$$

as well. In a similar way, demonstrating that the $(\gamma^{i})^2 = -I$, for $i = 1, 2, 3$, is also straightforward so we see that the **Pauli** matrix algebra **guarantees** the validity of the **Clifford** algebra. Furthermore, in Chap. 6 we investigate the algebra of the ***Quaternions*** which, it turns out, is really a ***special case*** of the Clifford algebra [34]. Thus the Pauli matrices also guarantee the validity of quaternion algebra.

12

Magnetic Moments

The magnetic dipole moment of the electron is the product of its spin and what is known as its gyromagnetic ratio. To explain the results of their historic experiments, Goudsmit and Uhlenbeck [34] found it necessary to assign the unusual value of 2 to that quantity (we recall that it was the subsequent need to refine that number which led to the development of QED). Pursuing the implications of the Dirac theory then provided a theoretical justification for the value of 2. This chapter summarizes a simple calculation first published in [10] that deals with an independent confirmation of that value for our alterative model and an associated item, the dipole magnetic moments of the nucleons. In what follows we shall dispense with the ABCD notation for a while and admit that we are talking here about the electron and the nucleons!

We begin with the nucleons for which, at the time of the writing alluded to above, the agreement of magnetic moment theory to experiment was not too good (see below). However, the ***ratio*** of the magnetic dipole moments for the two nucleons was in somewhat better shape which made it of interest to attempt an independent calculation in terms of our model. The approach was a strictly classical vector calculation as summarized in Figure 12-1.

The vectors indicated by the arrows represent the magnetic moments as being aligned with the spins of the individual quirks. Electric charge values are also indicated. As suggested by the figure we have taken advantage of the bilateral symmetry of electric charge for each nucleon. The ratio of magnetic dipole moments is then

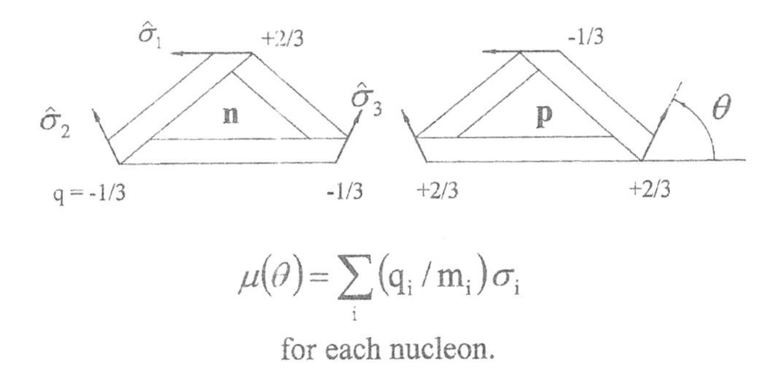

Figure 12-1. Nucleon magnetic moment ratio calculation.

calculated according to

$$R(\theta) = \frac{|\mu_n(\theta)|}{|\mu_p(\theta)|}, \tag{12-1}$$

where

$$\mu_k(\theta) = \sum_k \left(\frac{q_{ik}}{2m_{ik}c}\right)\sigma_{ik},$$

$i = n$ and p, k ranges over the three quirks, σ_{ik} is a unit quirk spin vector, the m_{ik} are the masses associated with the ith quirk and the quantity in parentheses is known as the gyromagnetic ratio. ***If*** these masses add up to the nucleon masses, Eq. (12-1) becomes

$$R(\theta) = \left(\frac{m_p}{m_n}\right) r(\theta), \tag{12-2}$$

where m_n and m_p are the masses of the neutron and proton, respectively, and $r(\theta)$ is found as the ratio of the magnitudes of the resultants of the vector diagrams of Figure 12-2.

Note that each spin vector has been weighted by the corresponding algebraic electric charge. From the vector diagram we find the ratio of the resultants to be

$$r(\theta) = 2\left\{\frac{[1+\sin^2\theta]}{[1+16\sin^2\theta]}\right\}^{1/2} \tag{12-3}$$

which, at $\theta = 90°$, has a minimum value of $\sqrt{8/17} = 0.685994$. Multiplying by the ratio of masses gives $R_{\text{Min}} = 0.685045$ which differs from the measured value of 0.684979 by 0.00964%.

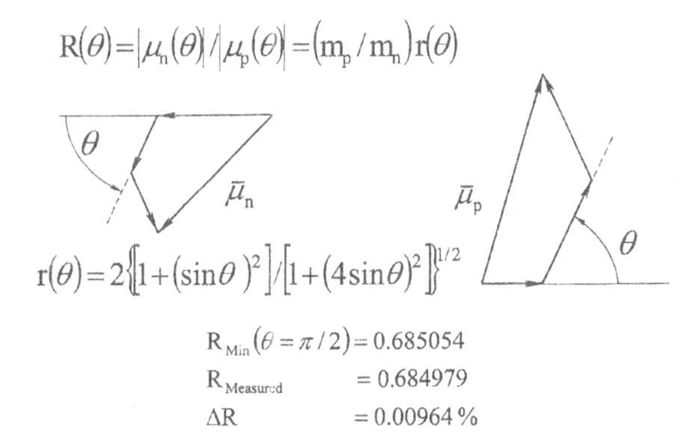

Figure 12-2. Moment ratio calculation continued.

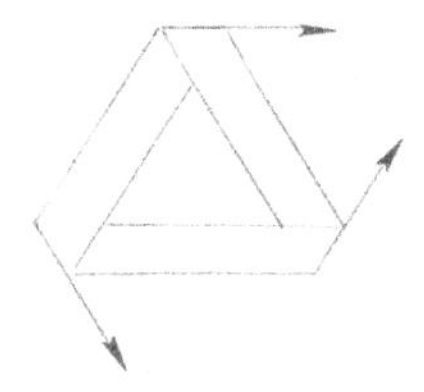

Figure 12-3. Electron magnetic moment vectors.

With regard to the electron, we find by the methodology described above that, given its equilateral symmetry as per Figure 12-3, its magnetic moment is

$$\mu_e = 2\left(\frac{e}{mc}\right)\sigma_q \tag{12-4}$$

rather than the expression given in connection with Eq. (12-1) for the nucleons. Here σ_q is the quirk spin and $2e/mc$ is the gyromagnetic ratio. The factor of 2, which confirms the Dirac theory arises simply as a result of vector addition.

The above has been criticized as not a quantum mechanical calculation but I do not see why that is necessary since our "particles" need not be viewed as quantum objects. Actually, I found a rather straightforward quantum mechanical calculation in an elementary particle textbook [36] and the answer differed from the experimental value by 2.76%, a factor of 286 times less accurate than the above.

13

Quaternions

Quaternions are useful for manipulating quantities in four dimensions (actually, in three as well) as are complex variables in two and we shall use them in Sec. VI. In this chapter we investigate how they might relate to the development of particle taxonomy, a very different point of view from the way we proceeded in Sec. II and illustrated with some diagrams you might find pertinent. A quaternion can be expressed as [32, 34, 37]

$$\mathbf{P} = p_0\lambda_0 + \mathbf{p}, \tag{13-1}$$

where

$$\mathbf{p} = p_1\lambda_1 + p_2\lambda_2 + p_3\lambda_3,$$

and

$$\begin{aligned} &\lambda_0\lambda_0 = 1, &\quad \lambda_1\lambda_2 = -\lambda_2\lambda_1 = \lambda_3, \\ &\lambda_0\lambda_i = \lambda_i\lambda_0 = \lambda_i, &\quad \lambda_2\lambda_3 = -\lambda_3\lambda_2 = \lambda_1, \\ &\lambda_i\lambda_i = -1, &\quad \lambda_3\lambda_1 = -\lambda_1\lambda_3 = \lambda_2. \end{aligned}$$

While the ps and qs provide the information explicit to diverse situations, the λs define the quaternion algebra common to all. Also, comparing their definition to that of the Clifford algebra, it is straightforward to show that the quaternion algebra is, indeed, a special case of Clifford algebra so that the Pauli algebra covers them both.

In any event, using the definition of the λs, we find the outer product of two quaternions to be expressible as

$$\mathbf{P} \otimes \mathbf{Q} = (p_0 q_0 - \mathbf{p} \cdot \mathbf{q}) + (p_0 \mathbf{q} + q_0 \mathbf{p}) + \mathbf{p} \times \mathbf{q}, \tag{13-2}$$

where

$$\mathbf{p} \cdot \mathbf{q} = p_1 q_1 + p_2 q_2 + p_3 q_3$$

and

$$\mathbf{p} \times \mathbf{q} = (p_2 q_3 - p_3 q_2)\lambda_1 + (p_3 q_1 - p_1 q_3)\lambda_2 + (p_1 q_2 - p_2 q_1)\lambda_3.$$

The outer product can of course also be expressed as a 4×4 matrix, the direct product of a column and a row vector (as in Chap. 3 to form boson matrix M) as

$$(p_0, p_1\lambda_1, p_2\lambda_2, p_3\lambda_3)^T \otimes (q_0, q_1\lambda_1, q_2\lambda_2, q_3\lambda_3)$$
$$= \begin{Vmatrix} p_0q_0 & p_0q_1 & p_0q_2 & p_0q_3 \\ p_1q_0 & -p_1q_1 & p_1q_2 & p_1q_3 \\ p_2q_0 & p_2q_1 & -p_2q_2 & p_2q_3 \\ p_3q_0 & p_3q_1 & p_3q_2 & -p_3q_3 \end{Vmatrix}. \tag{13-3}$$

However, we can also subdivide this matrix so as to summarize and highlight its intrinsic organization as a 2-dimensional manifestation of the content of Eq. (13-2). Thus, Figure 13-1, shows a ***scalar***, an ***inner product***, ***two vectors*** and a ***cross product***, the latter split between two pieces of matrix real estate, as shown in more detail together with the principal diagonal in Figure 13-2.

As discussed in the foregoing, the four basic FMSs give rise to a "super(anti)symmetric" kind of duplicate taxonomy: there is a half

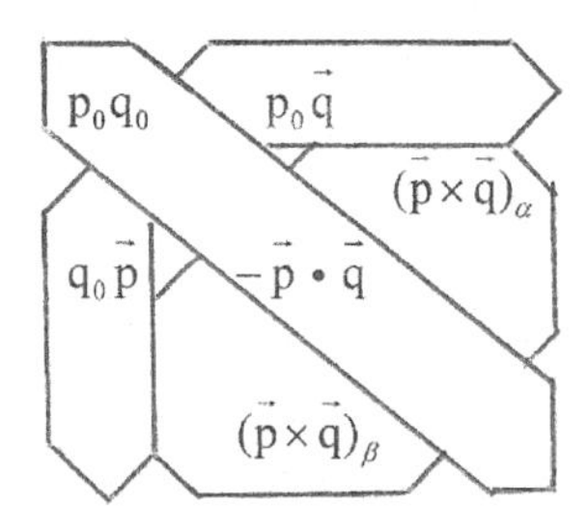

Figure 13-1. Quaternion product breakout.

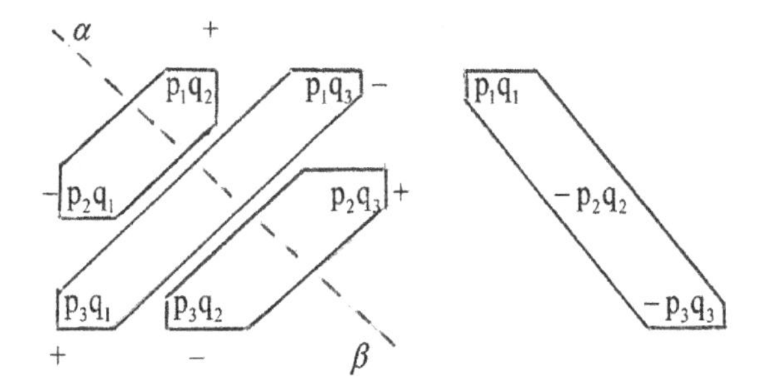

Figure 13-2. Inner and outer product breakout.

that involves the fermion triplet A, B and C and their conjugates and combinations and which largely replicates the taxonomony of the Standard Model; and there is another half involving D, C and B, which has no such connection but which exists by reasons of symmetry (see Sec. II). We recall that with the replacement of A, B and C by D, C and B, respectively, the two halves are manifestly isomorphic.

In Chap. 10 we have seen the development of a ***complex*** algebra as a result of the orthogonality of charge and twist in taxonomical development. Since (as per Sec. II), the taxonomy of the alternative model develops as per the dictates of the gauge group SU(2), it would appear that ***either*** of the above triplets ought to be amenable to treatment in terms of a ***quaternionic*** algebra, given the close association of that algebra and that group. The intent, here is to see if additional insights accrue thereby, beyond the previous treatment. Suppose, then, that as in the preceding we regard the three basic fermions A, B and C as a vector — that is as a pure quaternion — add an identity element and define a conjugate. Thus we consider as quaternions the 4-vectors

$$\begin{aligned} \boldsymbol{\Phi} &= \Phi_0 + \mathbf{F}, \\ \boldsymbol{\Phi}^* &= \Phi_0^* + \mathbf{F}^*, \end{aligned} \tag{13-4}$$

where

$$\begin{aligned} \Phi_0 &= f \\ \mathbf{F} &= \mathrm{A}\widehat{u_1} + \mathrm{B}\widehat{u_2} + \mathrm{C}\widehat{u_3} \end{aligned}$$

and similarly for the conjugate with the us being isomorphic to the λs. We remark that the unit quaternion is isomorphic to the gauge group SU(2) with the unit vectors being isomorphic to the Pauli matrices.

In the case of first order fusion, instead of Figure 13-1 we then have the matrix version of the outer product

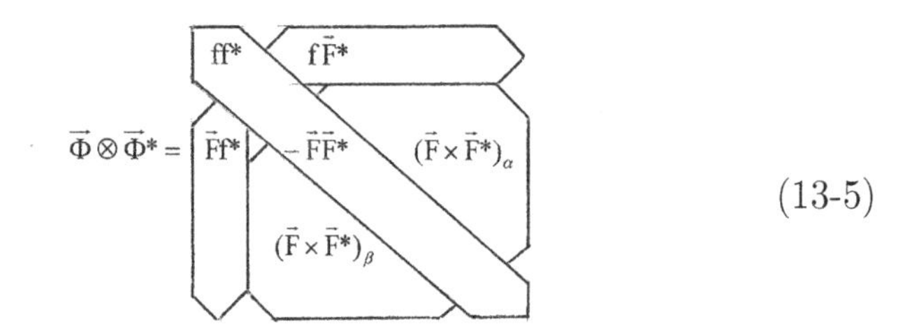

(13-5)

where

$$\overline{\mathrm{F}} \cdot \mathbf{F}^* = (\mathrm{AA}^*, \mathrm{BB}^*, \mathrm{CC}^*),$$

and

$$\mathbf{F} \times \mathbf{F} = (\mathrm{BC}^* - \mathrm{CB}^*)\widehat{u_1} + (\mathrm{CA}^* - \mathrm{AC}^*)\widehat{u_2} + (\mathrm{AB}^* - \mathrm{BA}^*)\widehat{u_3}.$$

Again, if we delete the scalar and vector terms (which involve identity elements f and f^*), we end up with a 3×3 matrix as in Figure 13-3. This is matrix M but rotated 90°. CW and without the row and column associated with enigmatic label D. It is organized according to the taxonomical analogs to Figure 13-2 for cross product and principal diagonal as shown in Figure 13-4.

$$\begin{Vmatrix} \mathrm{AA}^* & \mathrm{AB}^* & \mathrm{AC}^* \\ \mathrm{BA}^* & \mathrm{BB}^* & \mathrm{BC}^* \\ \mathrm{CA}^* & \mathrm{CB}^* & \mathrm{CC}^* \end{Vmatrix}$$

Figure 13-3.

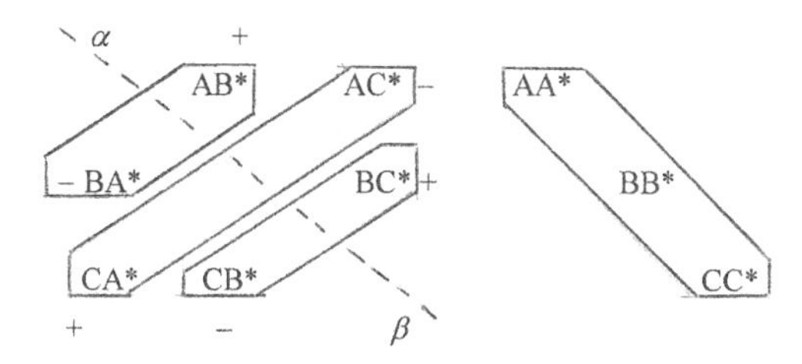

Figure 13-4. Inner and outer product breakout: First order fusion matrix.

In the case of second order fusion, we operate on a third fermion, say

$$\Theta = g + \mathbf{G} \tag{13-6}$$

with matrix $\Phi \otimes \Phi^*$, i.e. we form

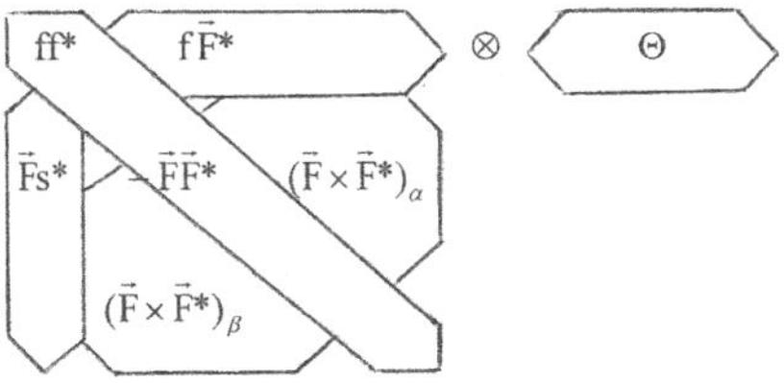

$(\vec{\Phi} \otimes \vec{\Phi}^*) \otimes \Theta =$ (13-7)

ff* $f\vec{F}^*$ $\otimes$ Θ

$\vec{F}s^*$ $\vec{F}\vec{F}^*$ $(\vec{F} \times \vec{F}^*)_\alpha$

$(\vec{F} \times \vec{F}^*)_\beta$

However, since we are only interested in the case where $f = f^* = g = 0$, this reduces, using Figure 13-4 and Eq. (13-5), to the direct sum of two outer products,

$$\left\| \begin{matrix} AA^* & 0 & 0 \\ 0 & BB^* & 0 \\ 0 & 0 & CC^* \end{matrix} \right\| \otimes (A, B, C)$$

and

$$\left\| \begin{matrix} 0 & AB^* & AC^* \\ BA^* & 0 & BC^* \\ CA^* & CB^* & 0 \end{matrix} \right\| \otimes (A, B, C)$$

which evaluates as the direct sum of six matrices, listed below, which together contain a total of 27 three-letter terms, each of which coincides uniquely with one of the 27 three-letter terms that remain in Figure 5-5 of Sec. II, after all terms containing letters D or D* are deleted

$$\left\| \begin{matrix} 0 & 0 & 0 \\ 0 & 0 & 0 \\ CC^*A & CC^*B & CC^*C \end{matrix} \right\|, \quad \left\| \begin{matrix} 0 & 0 & 0 \\ BB^*A & BB^*B & BB^*C \\ 0 & 0 & 0 \end{matrix} \right\|,$$

$$\begin{Vmatrix} AA^*A & AA^*B & AA^*C \\ 0 & 0 & 0 \\ 0 & 0 & 0 \end{Vmatrix}, \quad \begin{Vmatrix} AC^*A & AC^*B & AC^*C \\ BC^*A & BC^*B & BC^*C \\ 0 & 0 & 0 \end{Vmatrix},$$

$$\begin{Vmatrix} AB^*A & AB^*B & AB^*C \\ 0 & 0 & 0 \\ CB^*A & CB^*B & CB^*C \end{Vmatrix}, \quad \begin{Vmatrix} 0 & 0 & 0 \\ BA^*A & BA^*B & BA^*C \\ BA^*A & BA^*B & BA^*C \end{Vmatrix}.$$

Collecting terms then reduces the direct sum to the outer product of two vectors

$$\begin{pmatrix} A\sigma^* \\ B\sigma^* \\ C\sigma^* \end{pmatrix} \otimes (A, B, C),$$

where σ^* is the direct sum of three terms,

$$\sigma^* = (A^* \oplus B^* \oplus C^*).$$

This expression still encompasses all of the 27 terms but it is certainly more succinct as well as quite different from the formalism of Sec. II! Other than coming up with a rather neat, alternative way to summarize the taxonomy, I am not exactly sure how to evaluate this chapter. However, it provides an introduction to quaternions which we shall meet again in the discussion of time in Sec. VI.

14

A Variant of the Bracket for (2, n) Torus Knots

Before we make an explicit connection to the particles of the Standard Model, here is a topic that has to do with the algebra of (2, n) torus knots, specifically with connections between different values of the parameter n. How it might find application will be discussed later but the relationship of Möbius strips (MS) to (2, n) torus knots or links is well-known and was discussed in some detail in the preceding, a key aspect being that the half-twist (NHT) of a given MS is equal to the n value of the associated (2, n) torus knot or link. Thus, citing the equivalence between, for example, a ***composite*** FMS, whose NHT is the sum of two individual NHTs and a knot whose n is a composite ***number***, the sum of the two associated individual n values, allows us to study the latter to learn about the former. When we get to some rather free-form ruminations in the cosmological area in Sec. VII, we shall invoke what this chapter talks about (forewarned is forearmed).

We begin with a well-known (and oft-cited) knot/link invariant with important connections to physics as well as knot theory, the ***Kauffman bracket*** polynomial, colloquially known as "the ***Bracket***" [26], whose development, we note, stemmed from certain statistical considerations (see below). Here, of course, we are concerned with (2, n) torus knots for which Kauffman's canonical recursion formula as presented in [26], is

$$K_n = AK_{n-1} + (-1)^{n-1}A^{2-3n}. \qquad (14\text{-}1)$$

Cultural Aside: {Generally speaking the dimensional nature of the A's need not be specified but in this case, as seen in Chap. 9, the A's are in fact the solution to the Yang–Baxter equation that emerges in Kauffman's development of the bracket beginning with the consideration of two-dimensional arrays such as two-valued spin networks which leads to his analog in terms of the crossings encountered in traversing the usual 2-dimensional display of a knot. It is interesting that our FMSs constitute a very simple manifestation of such arrays, in this case of ***quirks*** which are either "up" or "down".}

The associated closed form version is readily shown to be

$$K_n = A^n K_0 + \sum_{m=1}^{n} (-1)^{m-1} A^{2-3m} A^{n-m}, \quad n \geq 1$$
$$= A^n K_0 - A^2 \sum_{m=1}^{n} (-A^{-3})^m A^{n-m}, \tag{14-2}$$

where

$$K_0 = -(A^2 + A^{-2}).$$

Basically, the summation here is in the form of a convolution (a symbolic convolution) but, it is helpful to use the following more conventional formulation,

$$K_n = A^n K_0 - A^2 \sum_{m=1}^{n} x^m y^{n-m}. \tag{14-3}$$

With this notation Eq. (14-2) becomes

$$K_n = AK_{n-1} - A^2 x^n. \tag{14-4}$$

Although a convolutional approach to discussing FMS composition was employed at length in Sec. II, here we adopt a rather different point of view, noting that the form of these equations suggests an isomorphic relationship to the field of sampled-data feedback systems as summarized in the diagram of Figure 13-1. (The open switch in the figure signifies that the K_0 input exists for only an initial sample.)

As in the subject of fusion emphasized in the foregoing, we are also interested in the case of composite n for $(2, n)$ torus knots, that is, an n that can be expressed as the sum of two or more components. The

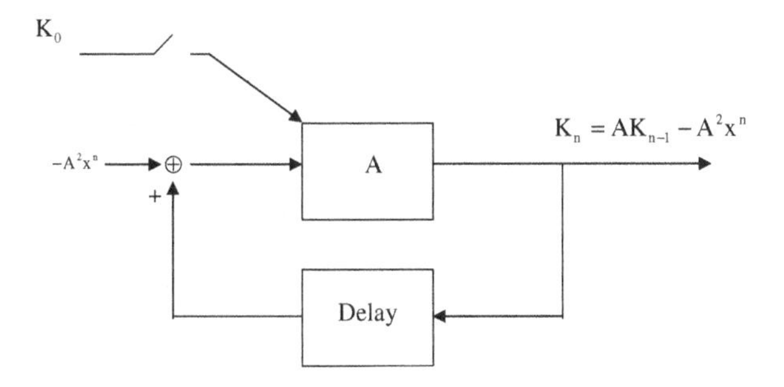

Figure 14-1. Schematic of (2, n) torus knot convolution.

simplest such example is the case of two components, i.e. $n = n_1 + n_2$ with both terms on the R.H.S. being positive. In which case we find

$$K_n = A^n K_0 - A^2 \left\{ y^{n_2} \sum_{m=1}^{n_1} x^m y^{n_1-m} + x^{n_1} \sum_{m=1}^{n_2} x^m y^{n_2-m} \right\}. \quad (14\text{-}5)$$

A more concise and symmetrical formulation is

$$K_n = A^n K_0 - A^2(\sigma_1 C_2 + \sigma_2 C_1)/2, \quad (14\text{-}6)$$
$$\sigma_i = x^{n_i} + y^{-n_i}$$

where

$$C_i = \sum_{m=1}^{n_i} x^m y^{n_i-m} \quad i = 1, 2$$

which looks like the weighted sum of two convolutions.

The case for three components proceeds in essentially the same way and in analogy with Eq. (14-6) we find

$$K_n = \prod_1^3 A^{n_i} K_0 - A^2(y_2 y_3 C_1 + y_3 x_1 C_2 + x_1 x_2 C_3), \quad (14\text{-}7)$$

where the coefficients are

$$\alpha_i = \alpha^{n_i}, \quad \text{for } \alpha = x, y, \ i = 1, 2 \text{ and } 3.$$

And the convolutions, C_i, are as before but with $i = 1, 2$ and 3.

Cyclic permutation of the subscripts gives two more equations and collecting terms associated with each of the three convolutions produces the final, symmetrical form,

$$K_n = \prod_1^3 A^{n_i} K_0 - A^2(\beta_1 C_1 + \beta_2 C_2 + \beta_3 C_3)/3, \tag{14-8}$$

where

$$\begin{aligned} \beta_1 &= y_2 y_3 + y_2 x_3 + x_2 x_3, \\ \beta_2 &= y_3 y_1 + y_3 x_1 + x_3 x_1, \\ \beta_3 &= y_1 y_2 + y_1 x_2 + x_1 x_2, \end{aligned}$$

which is the weighted sum of three convolutions.

We can readily generalize to the case of an arbitrary number, n_F, of components. In analogy with Eq. (14-8) we find that

$$K_n = \prod_1^{n_F} A^{n_i} K_0 - A^2(\gamma_1 C_1 + \cdots + \gamma_{n_F} C_{n_F})/n_F. \tag{14-9}$$

For any particular value of n_F, the coefficients (the γs) are computable in the manner illustrated for the case of $n_F = 3$. However, we can *anticipate* the results on the basis of the following selection rules:

- All subscripts, including that of the C's, must be present in each coefficient.
- The number of y terms decreases linearly from $n_F - 1$ in the first coefficient to 0 for the last, while the number of x terms increases, correspondingly, from 0 in the first coefficient to $n_F - 1$ in the last.
- The highest y subscript value is n_F and the highest x subscript value is $n_F - 1$.

For example, we can simply write down the equation analogous to Eq. (14-9), for the case of $n_F = 4$ as

$$\begin{aligned} K_n = {} & \prod_1^4 A^{n_i} K_0 \\ & - A^2(y_4 y_3 y_2 C_1 + y_4 y_3 x_1 C_2 + y_4 x_1 x_2 C_3 + x_1 x_2 x_3 C_4)/4. \end{aligned} \tag{14-10}$$

An easy way to view what is happening here is to consider four subscripts arrayed as 4 3 2 1 and a convolution indicator, C_i, that moves to the left from position 1 to position 4 as its subscript ranges from 1 to 4. At each position, each subscript to the *left* of the convolution indicator position has subscript value y and each subscript to its *right* has subscript value x. In effect the *convolution indicator* serves as a partition between the x set and the y set in each coefficient. For example, consider convolution indicator C_2: in this scheme, it is in the 2 position so we produce the term $y_4y_3C_2x_1$, or in the format of Eq. (13-10), $y_4y_3x_1C_2$. Basically, we have an invariance, the conservation of subscript values such that, in each coefficient, the sum of the x subscript values plus the y subscript values plus the value of the convolution indicator subscript must equal $n_F(n_F + 1)/2$ (for this example of course, it is equal to 10). For a given value of n_F, the implied *group structure* uniting the n_F coefficients in the analog to Eq. (14-10) is then simply that of *translation* over a set of contiguous n_F integers.

Here we supply some detail omitted from the above: To obtain Eq. (14-5) from Eq. (14-4) we split the summation into two consecutive summations as shown in Eq. (14-1), multiply and divide the second summation by x^{n_1} and relabel the $(m - n_1)$ term in that summation as m.

$$\sum_{1}^{n_1+n_2} x^m y^{n_1+n_2-m}$$

$$= y^{n_2} \sum_{1}^{n_1+n_2} x^m y^{n_1-m} + x^{n_1} \sum_{n_1+1}^{n_1+n_2} x^{m-n_1} y^{n_2-(m-n_1)}. \quad (14\text{-}11)$$

The case of $n_F = 3$ proceeds via the same kind of operations. As shown in Eq. (14-12) there are now three summations, two x factors to multiply and divide by and, correspondingly, two relabelings.

$$\sum_{1}^{n} x^m y^{n-m} = y^{n_2} y^{n_3} C_1 + y^{n_3} x^{n_1} C_2 + x^{n_1} x^{n_2} C_3. \quad (14\text{-}12)$$

Point of interest: It turns out that this convolutional version of the Bracket for (2, n) torus knots has found a use for itself in some rather speculative notions in the cosmological ruminations of Sec. VII!

IV
The Standard Model Connection

Well, finally! I hope the preceding subject matter made the delay worthwhile!

"*Patience is a virtue found to be quintessential in organizing a book that purports to show how elementary particles owe their attributes, indeed, their very existence to Knots, Braids, and Möbius strips*"

(Jack Avrin, 2014)

"*And probably of no small help in reading such a book, either.*"

(Potential Reader)

15

Ambiguities and Direct Comparison

Full disclosure: The main reason so much space was used to develop an alternative model using a general, lexicographic labeling for particles instead of going directly into standard notation is that our alternative model is inherently ambiguous; a given AM label (one of the capital letters) may represent more than one fundamental particle of the Standard Model depending on the *interaction* being modeled. The ambiguity is summarized in Figure 15-1.

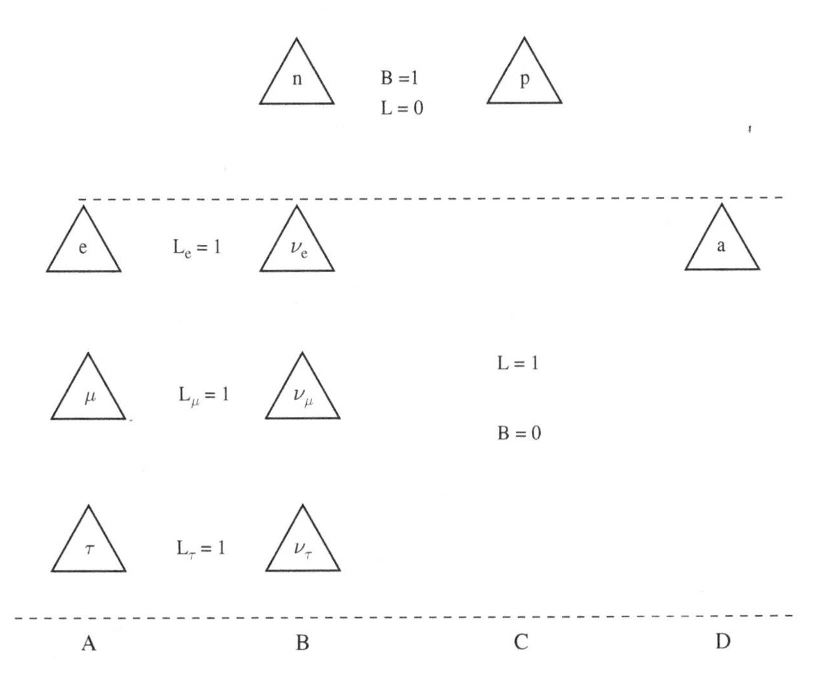

Figure 15-1. Standard model ambiguities in the alternative model system.

As indicated, the nucleons are represented in the top row and the leptons in the next three rows. The AM labels employed heretofore are in the bottom row and the idea of the figure is that, *a priori*, such a label may represent any one of the particles in its column. However, the label C represents the proton unambiguously (but read on) and the label D likewise represents only that enigmatic particle whose existence is required by symmetry (but whose role is problematical). On the other hand, there is a triple (leptonic) ambiguity associated with label A and a quadruple ambiguity associated with label B which can represent either the neutron or one of the three neutrinos.

About the proton: it might occur to some that there's a striking difference in the appearance of the neutron and the proton columns; in this alternative model the two basically differ just in the direction of their twist and in the Standard Model theory of isospin they are just two directions of a single vector. So, if the neutron column has ***neutrinos***, shouldn't the proton column have ***protinos***? The answer is no but we'll talk about it later in the book.

But, back to the matter of ambiguity, there's no cause for alarm! In practice, this kind of ambiguity is not really a problem; it can be resolved by invoking the rules that constrain which interactions are realizable. Some of these are derivable from basic principles (conservation of charge is one such) and some are not, for example, conservation of *baryon* and *lepton* numbers. Such ambiguity resolution will be illustrated presently in terms of well-known interactions, but in the meantime, we note that, in the *reverse* situation — that is, where it is the Standard Model elements that are *specified* to begin with — the corresponding Alternative Model elements are indicated *unambiguously*.

In any event we now exhibit the AM/SM comparison in terms of the bosons that result from first-order fusion, first the AM's boson matrix repeated once more:

$$M_{\mathrm{AM}} = \begin{bmatrix} - & \mathrm{BD}^* & \mathrm{CD}^* & \mathrm{DD}^* \\ \mathrm{AC}^* & \mathrm{BC}^* & \mathrm{CC}^* & \mathrm{CD}^* \\ \mathrm{AB}^* & \mathrm{BB}^* & \mathrm{CB}^* & \mathrm{DC}^* \\ \mathrm{AA}^* & \mathrm{BA}^* & \mathrm{CA}^* & - \end{bmatrix}$$

and then translated (per various references including [36] into SM nomenclature in Eq. (15-1), mainly according to the nomenclature of Figure 15-1 plus the role of the bosonic members of the matrix in the interactions we discuss below in Chaps. 16 and 17.

$$M_{\text{SM}} = \begin{bmatrix} — & (\nu_e a^*, na^*) & pa^* & aa^* \\ ep^* & \pi^- & \pi^{0\text{R}} & ap^* \\ W^- & (Z^0, \pi^{0\text{L}}) & \pi^+ & an^* \\ (\gamma, Z^0) & W^+ & pe^* & — \end{bmatrix}. \tag{15-1}$$

The empty corners here signify the previously mentioned impossible combinations, and in fact, the upper row and right-hand column — six items in all — are also combinations that are not involved in actual SM interactions just because they involve the enigmatic "a". A point of interest: the reader may recall the remark of Chap. 5 that matrix M can be regarded as an ***operator*** that converts a vector representing the set of four basic, single component (letter) fermions into a vector of matrices, one that codifies the set of three-letter words representing the set of three-component fermions. Of course, the operator role is actually *implemented* by the individual elements (in other words, the bosons) of the matrix. In detail, those operators involved in "electroweak" interactions, namely γ, W^-, W^+ and Z^0 are in the lower left quadrant of both versions. In the (diagonally overlapping) next quadrant are the ones involved in "strong" interactions: π^-, π^+, $\pi^{0\text{L}}$ and $\pi^{0\text{R}}$ and, all of a sudden we see another deviation; two versions of the neutral pion in the AM version. We will have more to say about that later but, according to [38], there is empirical evidence that the neutral pion does, in fact, come in two versions both of which participate in the Delta particle interactions shown below.

16

Some Interactions: Delta Creation and Decay

We now consider the creation of delta particles by the ***excitation*** of nucleons operated upon by ***pions*** (or conversely, delta decay into nucleons). Since no leptons are involved, these interactions are unambiguous so there is no reason not to express this directly in SM notation. However, for direct comparison purposes, we begin with our model's notation in the associated FMS schematic shown in Figure 16-1.

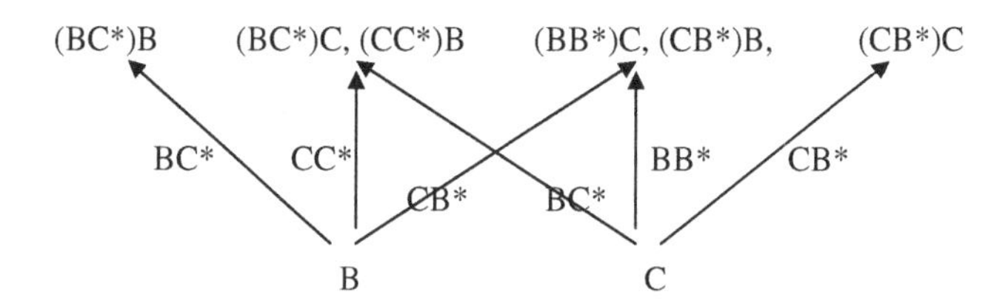

Figure 16-1. Delta particle creation and decay: AM notation.

Thus, we see component B (which translates as n) of basic fermion 4-vector $V = (\mathrm{A, B, C, D})^{\mathrm{T}}$ *transformed* into the component BC*B (translates as Δ^-) of the 4-vector P3 (See Chap. 5) by the operation of matrix element BC* of M, which as per Eq. (16-1) corresponds to the vector boson π^-, and also into component CB*B (translates as Δ^+) of vector P3 by matrix element CB* which corresponds to the vector boson π^+. There are similar interactions associated with component C (which translates as p) of vector V, the excited states Δ^0 and Δ^{++} (translating as components BC*C and CB*C, respectively, of vector P2) and, again, the vector bosons π^- and π^+.

In a sense, BC* (i.e. π^-) and CB* (i.e. π^+) both function as identity elements in that the *twist* of whatever MS they multiply (operate upon) is retained. The reason is that the twist of each is $n = 0$ because each is formed by applying a further twist, to the left or the right, to a *topologically untwisted* ***closed*** band. On the other hand, each adds (or subtracts) a unit of *charge* to the operated-upon vector component. Conversely, matrix elements BB* and CC* are identified in [1 and 2] with what is defined as π^{0L} and π^{0R}, respectively. Each has charge $q = 0$ and so functions as an identity element in that respect but adds (or subtracts) two units of *twist*. Thus, CC* operates upon B (i.e. n) to create CC*B (i.e. a Δ^0) and BB* operates upon C (i.e. p) to create BB*C (i.e. a Δ^+). The net result is summarized in the table below:

Word	BC*B	BC*C, CC*B	CB*B, BB*C	CB*C
Particle	Δ^-	Δ^0	Δ^+	Δ^{++}
Twist	−1	+1	−1	+1
Charge	−1	0	+1	+2

In summary, the figure portrays the deltas as excited states of either the neutron or the proton in a single way for the Δ^- and the Δ^{++} and in two ways for the Δ^0 and the Δ^+. Conversely, they can decay back into the nucleons in six ways, three for the neutron and three for the proton.

Figure 16-2 is the translation of Figure 16-1 into Standard Model notation (as per [39] and Figure 16-3 (borrowed from a presentation given at a symposium — I forget which — a long time ago!) shows the associated FMS schematic but with the SM notation.

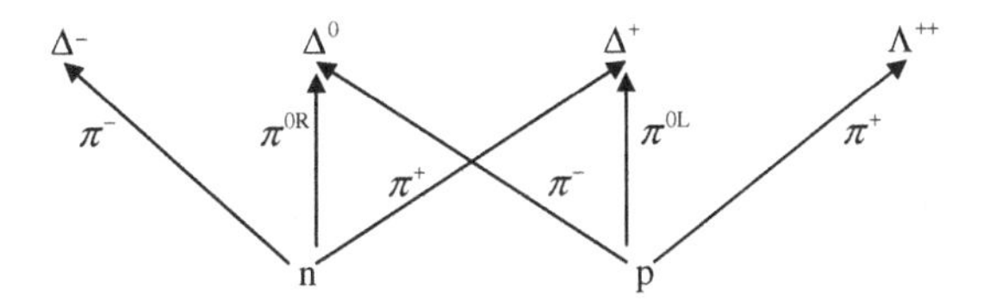

Figure 16-2. The same in SM notation.

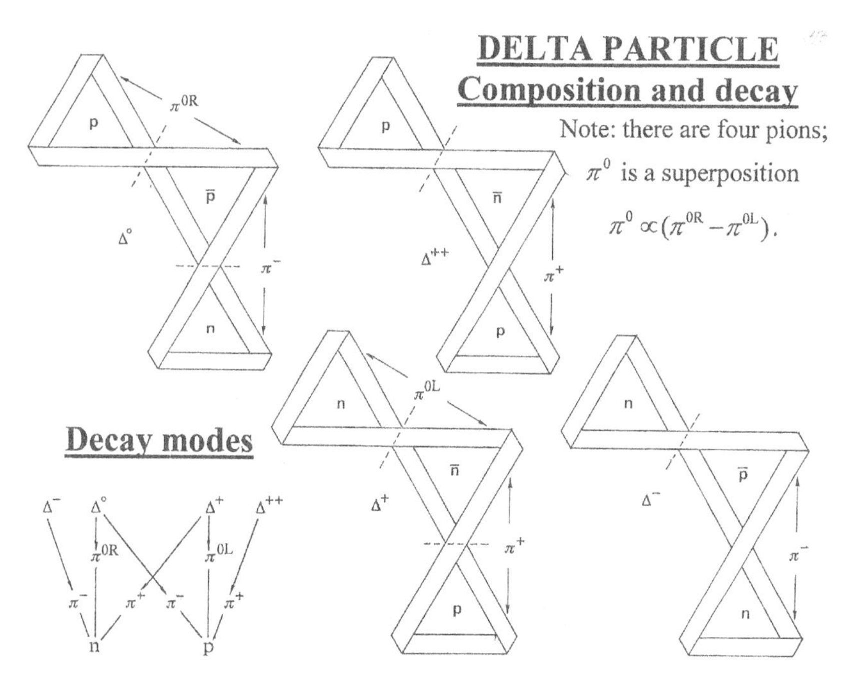

Figure 16-3. FMS schematic of delta creation and decay (SM notation).

To help with the translation we also reproduce below figures for the pions π^+ and π^- from Chap. 3 of Sec. II.

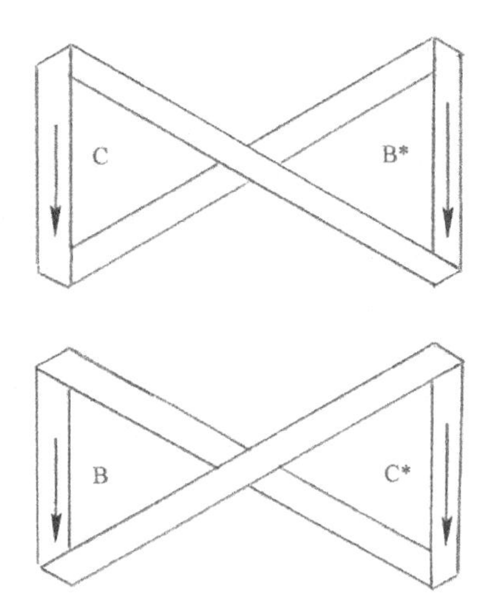

Charged pions (repeat).

That is, we have

$$\pi^{+} = \mathrm{CB}^{*} \quad \text{and} \quad \pi^{-} = \mathrm{BC}^{*}. \tag{16-1}$$

In summary, viewing these pions as mesons operating on the nucleons as per previous discussion, the delta particles are

$$\begin{aligned} \Delta^{-} &= \mathrm{BC}^{*}\mathrm{B} \leftrightarrow \pi^{-}n, \\ \Delta^{0} &= \mathrm{BC}^{*}\mathrm{C} \leftrightarrow \pi^{-}p \ (\text{or } \pi^{0\mathrm{R}}n), \\ \Delta^{+} &= \mathrm{CB}^{*}\mathrm{B} \leftrightarrow \pi^{+}n \ (\text{or } \pi^{0\mathrm{L}}p), \\ \Delta^{++} &= \mathrm{CB}^{*}\mathrm{C} \leftrightarrow \pi^{+}p. \end{aligned} \tag{16-2}$$

In other words, as modeled in the AM, our Δ^{++} and Δ^{0} are seen to be excited states of the *proton* and similarly, Δ^{+} and Δ^{-} are excited states of the *neutron*, as they should be per Figures 16-1 and 16-2. Correspondingly, our Deltas form two Isospin pairs, namely (Δ^{-}, Δ^{0}) and (Δ^{+}, Δ^{++}) which, as per the SM (e.g. [39]) they should.

About the duplication of the SM's neutral pion: as indicated in Figure 16-3 we can establish a correspondence between the latter and our two alternate model versions by defining the superposition

$$\Pi^{0} = \pi^{0\mathrm{R}} - \pi^{0\mathrm{L}} = \mathrm{uu^{*}u^{*}u} - \mathrm{dd^{*}u^{*}u}. \tag{16-3}$$

Upon eliminating the common factor, $\mathrm{u^{*}u}$ from each term we are left with

$$\Pi^{0} \propto \mathrm{uu^{*}} - \mathrm{dd^{*}}, \tag{16-4}$$

which "is the accepted SM composition (also viewed as a *superposition*) for the π^{0}" [10]. In the same vein, we note that the two pion versions can also form the superposition for the η particle, viz:

$$\eta \propto \pi^{0\mathrm{R}} + \pi^{0\mathrm{L}} \rightarrow \mathrm{uu^{*}} + \mathrm{dd^{*}}. \tag{16-5}$$

17

Beta Decay

In the case of Delta creation and decay, translation between the two notations was unambiguous. However, in modeling weak interactions that is no longer the case. In any event, we begin with neutron decay, our original notation and with the excited states shown in the NHT $= -1$ column of Figure 5-5, noting specifically the terms AC*C and CC*A (which are operationally identical for our purpose) in the uppermost triplet of the NHT $= -1$ column. Using CC*A, the stick figure schematic of Figure 17-1 below then shows it as the first stage, a folded-over (excited) version of the original FMS configuration of B, the Alternative Model's surrogate for the neutron, n. On the basis of the quirk structure, C and C* translate unambiguously to p and p*, which conserves twist but ***not*** *baryon* number or *charge*.

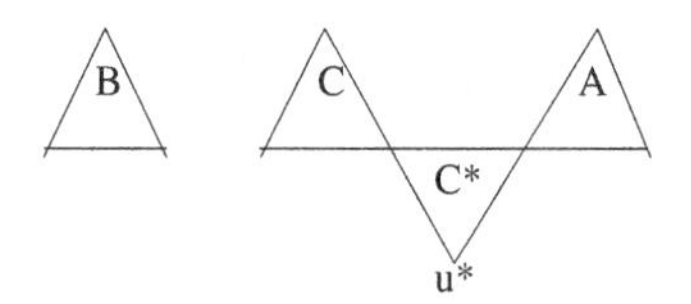

Figure 17-1. First stage of neutron decay (AM notation).

Looking ahead a bit, what we want is the Alternative Model decay process, B $\to$ C + B* + A, which translates to the SM decay [40], $n \to p + \nu^* + e$. We can identify C with the proton but there's a caveat about C*; it would be too heavy to translate as an (anti) proton. A tentative solution; it's the protino we just talked about but in the conjugate form.

Thus, we need to increase the baryon number in the diagram by one and cancel the lepton number and charge of A. Furthermore, on the basis of *mass*, A should translate to e rather than μ or τ (which are too heavy) in which case we need to change the C* to something that translates to the corresponding antineutrino, ${\nu_e}^*$ (for its lepton number).

All this can be accomplished if we just change the indicated u^* *quirk* into a d^* *quirk*! In [10] this is referred to as the "***beta switch***" and it is further clarified in knot theoretic terms here in Appendix E. Thus, here we go from CC*A, an excited state of the neutron, to CB*A, an excited state of the electron. In terms of twist, we go from NHT = −3 to NHT = −1 by changing the u* quirk with NHT = +1 to the d* quirk with NHT = −1. The meson operator is B*A, the Alternative Model version of the W^- particle (see the boson matrix M, Eq. (15-1) which then splits into an *electron* and its *antineutrino* as indicated in Figure 17-2.

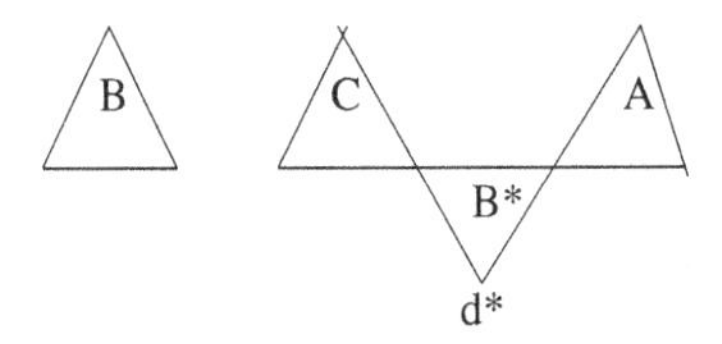

Figure 17-2. Second stage of neutron decay (AM notation).

Translation into the corresponding SM notation is shown in Figure 17-3 (also borrowed from the cited presentation). Note the analogy to the Standard Model's ***unexplained*** switch of one of the neutron's down quarks to an up quark, by means of which the neutron is converted into a proton plus a W particle.

Muon decay is a similar process; we begin again with an *excited state*, this time in the NHT = −3 column of Figure 5-5, specifically noting the terms AC*B and, equivalently, BC*A of the sextet. Using BC*A, we see in Figure 17-4 a diagram very similar to Figure 17-1 and it goes through an identical beta switch process, leading to a similar diagram in Figure 17-5.

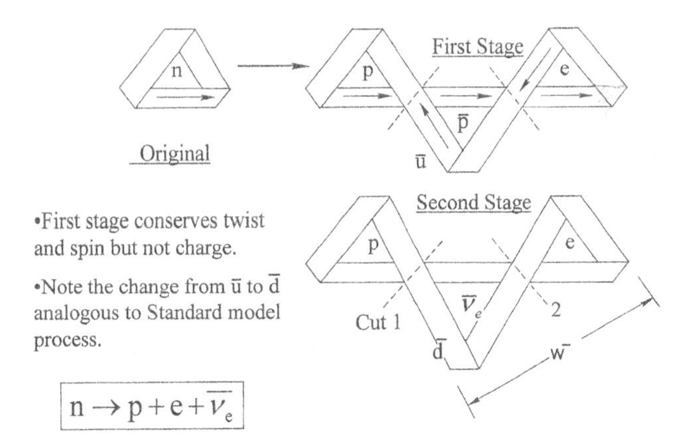

Figure 17-3. FMS neutron decay schematic; SM notation.

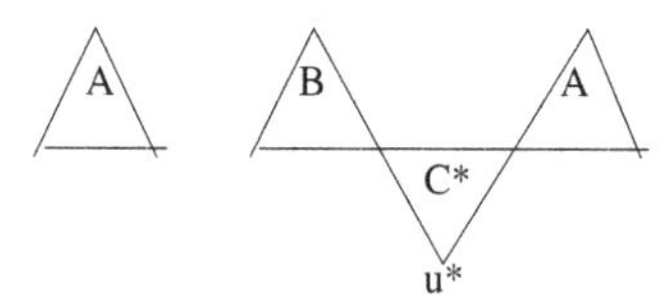

Figure 17-4. First stage muon decay; AM notation.

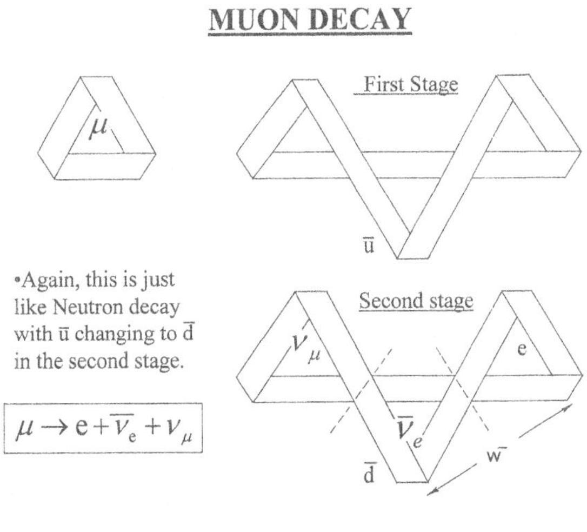

MUON DECAY

First Stage

μ

$\bar{u}$

•Again, this is just like Neutron decay with $\bar{u}$ changing to $\bar{d}$ in the second stage.

$\mu \rightarrow e + \overline{\nu_e} + \nu_\mu$

Second stage

ν_μ

e

$\bar{\nu}_e$

$\bar{d}$

w^-

Figure 17-5. FMS muon decay schematic; SM notation.

The ambiguity resolution logic of the foregoing is that, on the basis of mass, the A on the right can only translate to an e and, similarly, B must be a neutrino rather than a neutron. Again, there is an apparent baryon number and charge mismatch across the equal sign due to the C* which translates to a p^* (which, again, can only be the transitory antiprotino) and, again, it is corrected by the "*beta switch*" which changes the C* into a B*. In turn, that B* translates as an antineutrino which must be a ν_e in order to cancel the electron's lepton number. And finally, the B of the second fermion must translate as a muon neutrino ν_μ in order to match lepton numbers, whereupon we end up with BB*A which translates as $\nu_\mu W^-$ which splits up into an electron and its antineutrino and the muon's neutrino.

A final comment about beta decay modeling: there is an existential difference between the Alternative Model's beta switch and the corresponding SM model. Although beta decay in the SM ***looks*** simple diagrammatically; it depends on what we consider to be the mysterious conversion of a down quark into an up quark and a W vector boson; quite inexplicable in ontological terms. On the other hand, the beta switch of the AM can be simply rationalized in straightforward knot theoretic terms. What it does is change the up antiquirk into a down antiquirk, a change in ***twist*** which is discussed as the ribbon equivalent of the two-strand crossover switch of knot theory [26]. In Appendix D, "*beta switch*) subtraction" of the edge-on representations of the two antiquirks is seen to produce a representation of the ***unknot*** which on attachment to any knot configuration, has no effect and can therefore be dispensed with.

Interaction Recap:

*The **strong** interaction* involves one or more of the following:

Fusion of a nucleon and a pion to produce an excited state of the nucleon.
Fission of a nucleon and pion combination.
Fusion of a nucleon and an antinucleon to produce a pion.
Fission of a pion to produce a nucleon and an antinucleon.

Weak *decay* involves the ***folding*** of a fully formed nucleon in violation of charge and parity conservation.
Transformation of an antidown quirk into an antiup quirk regaining charge conservation and parity conservation.
Fission of the transformed excited state into a "partner" particle[1] and a vector boson.
Fission of the vector boson into an electron and antineutrino.

Folding is equivalent to a Reidemeister move in knot theory [26] which preserves twist. All excited states can be created by folding as well as fusion.

Consequently, the Delta particles replicate nucleon twist, being $+1$ for the pair (Δ^0 and Δ^{++}), and -1 for the pair (Δ^- and Δ^+), respectively.

[1]Proton in neutron decay and electron in muon decay.

18

“Deuteronomy” and Isospin Invariance

In [11] a model was shown of pions mediating Yukawa type exchanges between nucleons to maintain deuteron stability in what may be characterized as (strong) isospin space and we reproduce it here in Figure 18-1 (again from the cited presentation). This is a dynamic process, as summarized in Figure 18-1, (another presentation figure) postulated to maintain that stability. The figure is pretty much self-explanatory: all four pions of the AM are involved, two for the proton and two for the neutron. At each stage of the process what was a free proton becomes a neutron and conversely what was a free neutron becomes a proton. Also, in (what we might arbitrarily call the first stage) what was a π^- becomes a π^{0R} and what was a π^+ becomes a π^{0L}. Two fusions and two fissions take place, in each case. The process then reverses to recover the original pair of nucleons.

More generally, there is an important *functional* symmetry between the lower left-hand quadrant and the overlapping next diagonal quadrant of the boson matrix M: the manner in which the ***vector bosons*** in the first quadrant act upon the electron/(anti)neutrino pair of ***leptons*** (in what's known as ***weak*** isospin space) is *identical* to the manner in which the ***pions*** in the second quadrant act upon the ***nucleons*** (in what's known as ***strong*** isospin space). In Figure 18-2 we show a more abstract model that fits ***both*** strong and weak isospin cases according to the identifications as listed below and the table that follows then shows the specific identifications to the two cases.

DEUTERONOMY:
Pion exchange between Nuclei

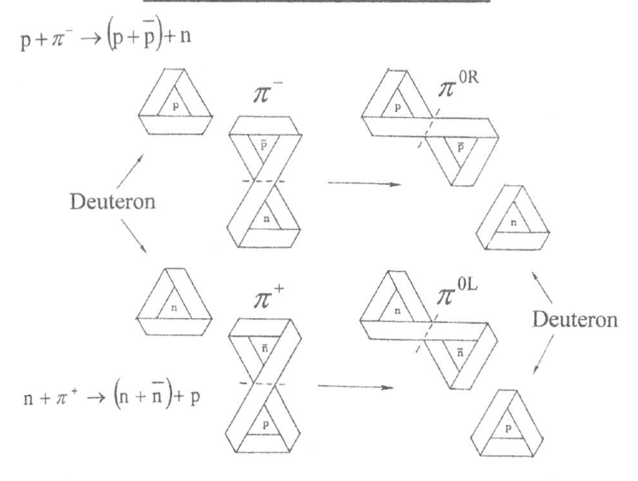

Figure 18-1. Deuteron stability as a dynamic process.

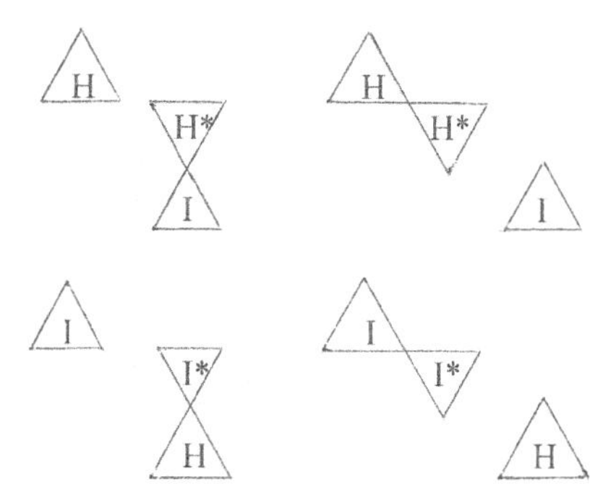

Figure 18-2. Generalized isospin manipulation.

Thus in this process, the π^- splits into a free neutron and an antiproton which fuses with the original proton to make π^{0R} and, similarly, the π^+ splits into a free proton and an antineutron which fuses with the neutron to form π^{0L}. That is we have the interactions

$$p + \pi^- \rightarrow (p + p^*) + n$$
$$n + \pi^+ \rightarrow (n + n^*) + p$$

Table 18-1: "Strong" and "Weak" isospin correspondence.

Strong		Weak
P	H	e
n	I	ν^e
π^-	J	W^+
π^+	K	W^-
π^{0R}	L	Z^{0L}
π^{0L}	M	Z^{0R}

such that we always have a free proton and neutron pair — basically what was presented above to illustrate the deuteron stability mechanism. But, now, if we make the correspondences indicated above between the *proton* and the *electron*, and also between the *neutron* and the *neutrino*, and so on, we can construct the ***same*** diagram for the stability of the electron/neutrino pair, with the W^+ splitting in analogy with the π^- and so on to perpetuate the electron/antineutrino pair in accordance with the interactions

$$e + W^+ \to (e + e^*) + \nu,$$
$$\nu + W^- \to (\nu + \nu^*) + e.$$

19

Simple Symmetry and CPT Invariance

The symmetry properties of the basic fermions are simple if there are no external influences. Of course the diagram for the electron (e) must be invariant to rotations about an axis normal to the plane of the diagram and, provided it is compensated by parity (direction of traverse) reversal to rotations of 180° about each of the axes a, b, and c as suggested in Figure 19-1.

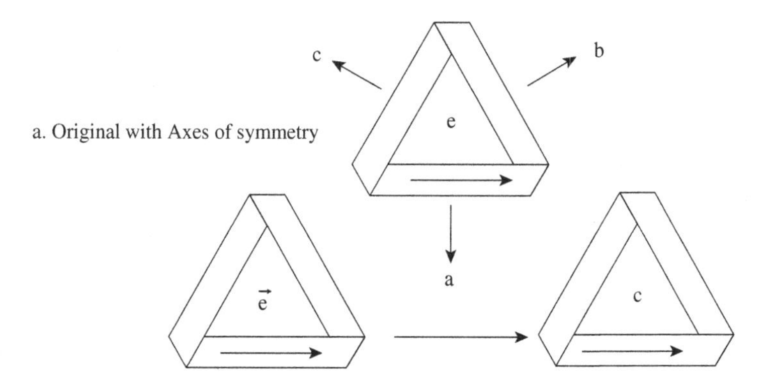

Figure 19-1. Symmetry properties of the Electron Diagram.

The situation for the nucleons is similar, the main difference arising as a result of the bilateral rather than equilateral particle symmetry. Here, rotations about the axes a, b, or c, again compensated by parity reversal, just form a set of three manifestation of the original, related by 120° rotations in the plane as indicated in Figure 19-2.

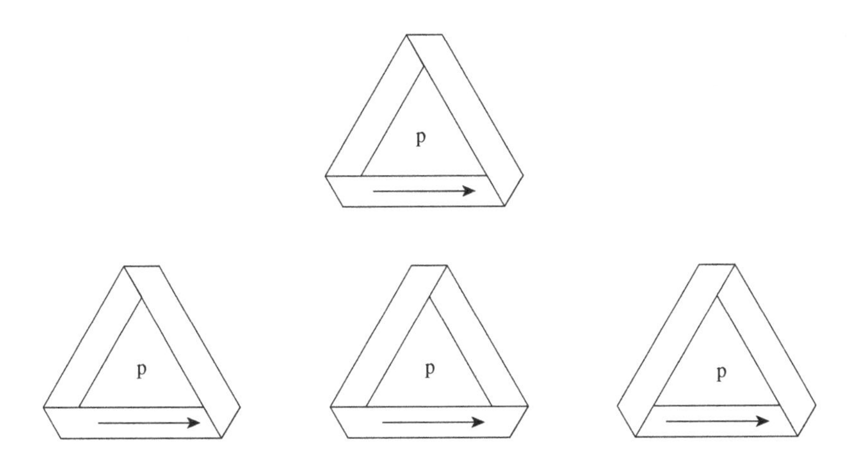

Figure 19-2. Proton revolution about axes a, b and c followed by parity reversal.

First order fusion is a bit more complex but it also offers a good opportunity to exhibit CPT invariance. In Figure 19-3, a generic first order fusion diagram leads to an associated list of possible combinations for the four quirks: Then, with the crossover labeled as either + or −, we can label orientations as shown in Figure 19-4 for the two W particles, where the x axis is pointed out of the plane. The

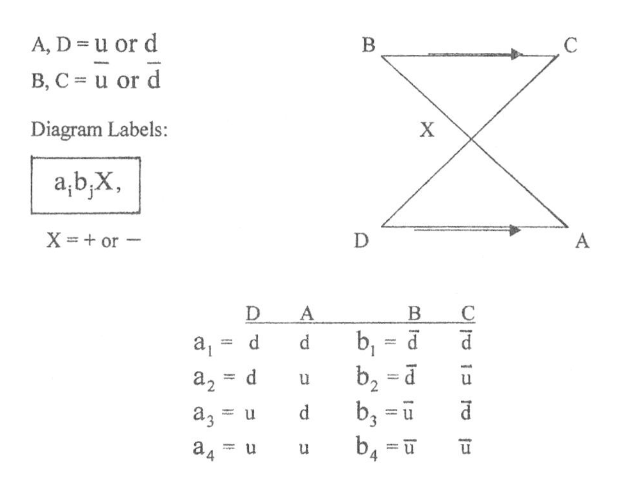

	D	A		B	C
a_1 =	d	d	b_1 =	$\bar{d}$	$\bar{d}$
a_2 =	d	u	b_2 =	$\bar{d}$	$\bar{u}$
a_3 =	u	d	b_3 =	$\bar{u}$	$\bar{d}$
a_4 =	u	u	b_4 =	$\bar{u}$	$\bar{u}$

Figure 19-3. Combinations for four quirks.

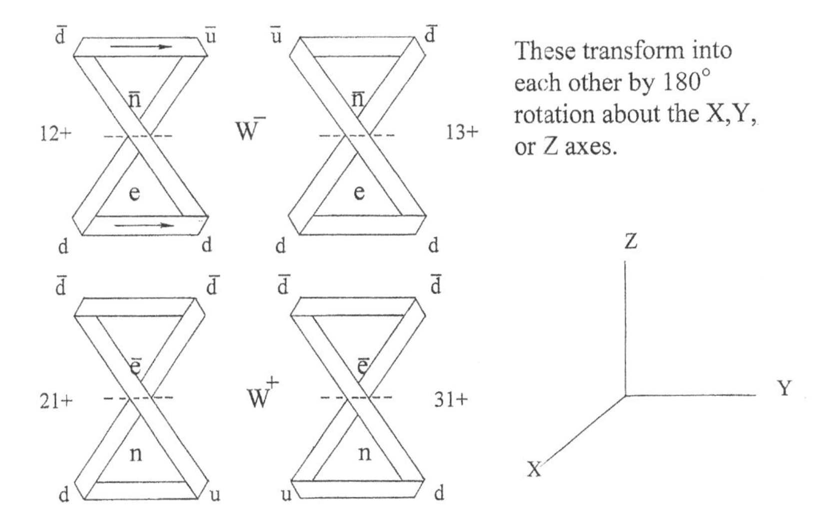

Figure 19-4. *W* particle orientations.

result of rotations about each of the three axes is then summarized in Figure 19-5, i.e. the 12+ version of W^- transforms into its 13+ version by a rotation about the z-axis and so forth.

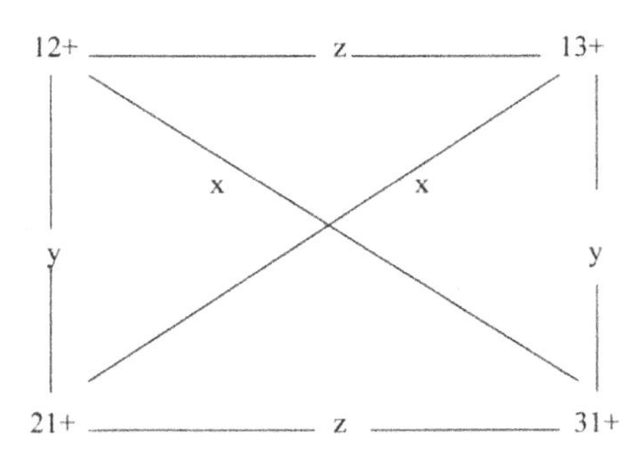

Figure 19-5. Result of 180° rotations.

Now if, as per the foregoing discussion, we associate the x-axis with time and the labels for Charge, Parity and Time as C, P, and T, respectively and their reversal as C*, P* and T*, respectively, rotations about each of the three axes are found to produce label transformations as follows:

$$@\ \mathrm{X} \Rightarrow \begin{pmatrix} \mathrm{C} \rightarrow \mathrm{C} \\ \mathrm{P} \rightarrow \mathrm{P} \\ \mathrm{T} \rightarrow \mathrm{T} \end{pmatrix} \quad @\ \mathrm{Y} \Rightarrow \begin{pmatrix} \mathrm{C} \rightarrow \mathrm{C}^* \\ \mathrm{P} \rightarrow \mathrm{P} \\ \mathrm{T} \rightarrow \mathrm{T}^* \end{pmatrix} \quad @\ \mathrm{Z} \Rightarrow \begin{pmatrix} \mathrm{C} \rightarrow \mathrm{C} \\ \mathrm{P} \rightarrow \mathrm{P}^* \\ \mathrm{T} \rightarrow \mathrm{T}^* \end{pmatrix}.$$

Assigning + to each of C, P and T and – to each of their conjugated versions shows that the CPT product is invariant to orientation.

20

Family Matters

So far, everything we have discussed pertains to a single generation of the SM, which, as we know, actually consists of a family of three generations of particles. The object of this section is to display a corresponding triplication of our alternative model, in fact a corresponding *family* structure consisting of three and *only three* generations. Although most of the argument is demonstrated in some detail in [10], here, while omitting some of the detail, we shall go a bit further in terms of the symmetries inherent to our model.

As in [10], we first arrange the four *basic* fermion representations, in order of twist, as shown in Figure 20-1, recalling the progressive replacement of d quirks by u quirks, and the corresponding increase in twist from −3 to +3 and charge from −1 to +2. Even though it plays no part in either the taxonomy or interactions of the model as developed in the foregoing, we include fermion D for the sake of symmetry (as we have all along to one extent or another).

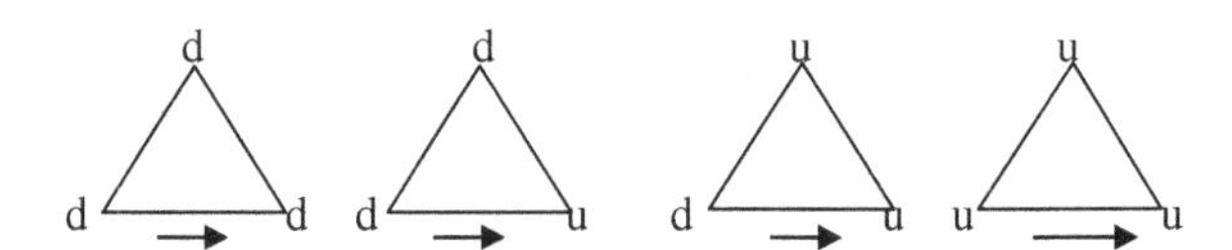

Figure 20-1. Four basic fermion schematic; Two-quirk version.

These four objects can also be viewed as four points on a line in what we define as "***label space***", shown in Figure 20-2 as a function of twist. The line is anchored by labels **D and U**, which, ***here***, stand for quirk structures of **ddd** and **uuu**, respectively. **Note**: In

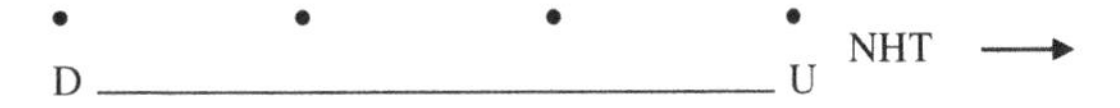

Figure 20-2. And their loci in twist space.

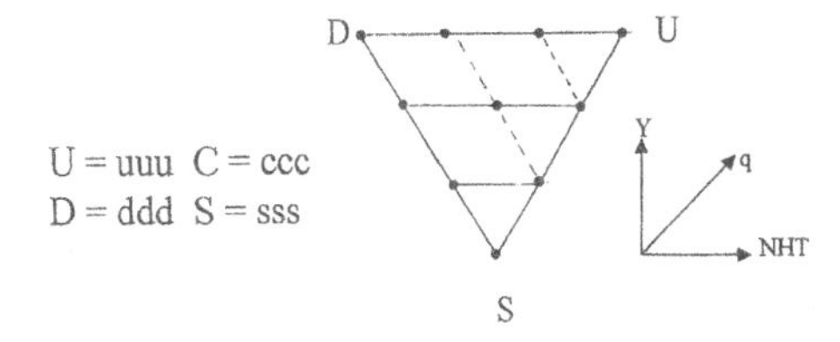

Figure 20-3. Three-quirk organization.

this section, D will ***no longer represent*** that enigmatic fermion with NHT = 3 and charge of +2. Please accept my apologies for the confusing switch in notation!

This representation initiates an organizing ***rationale*** for incorporating further generations. Thus when we consider an additional quirk, denoted by the letter **s** (in analogy with the SM "strange" quark), we just need to introduce its representation, capital letter S, in label space such that it interconnects with the other two labels. The result is the triangular array shown in Figure 20-3 which illustrates the replacement progression from a composition of three d quirks to one of three s quirks along a line anchored by capital letters D and S and similarly for the progression of replacing letter u by letter s.

Note that the triangular display accommodates ***seven*** appropriately placed intermediate points for a total of ***ten*** points, to each of which we can associate a particular member of a particular set of alternative model particles. In other words, we can view such a geometrical structure as a ***scaffold,*** the indicated points of which are loci upon which to emplace orderly arrangements of basic alternative model fermions or bosons, up to and including a *decuplet* arrangement. Note also, the ***isomorphic*** relationship of the DUS triangle to the *defining* representation of the gauge group SU(3), the triangular array featuring d, u and s quarks in the development of the SM's historic "*Eightfold way*".

Thus, Figure 20-3 can be considered as the AM's prototypical ***manifestation*** of a ***flavor*** SU(3) format which we will see replicated as we proceed to erect a family tree. The inclined dotted lines are loci of constant charge and the horizontal lines are hypercharge loci in each case so that twist, charge and hypercharge increase as shown in the accompanying coordinate system, all in direct analogy to the SM.

The key feature here is that of ***symmetrical interconnectivity*** of capital letter labels; if we impose that as a *requirement* for the incorporation of additional *flavors* it implies, to quote [10] "that the associated geometrical edifice must be *regular*. In particular, if there are N flavors, each flavored vertex must extend a line to $N-1$ distinct other flavored vertices. Thus adding another flavor, the (c) quirk, produces $4!/2!2! = 6$ lines, which combine in the ***tetrahedral*** structure" shown in Figure 20-4. Note that this structure also automatically subsumes $4!/3 = 4$ contiguous triangular arrays.

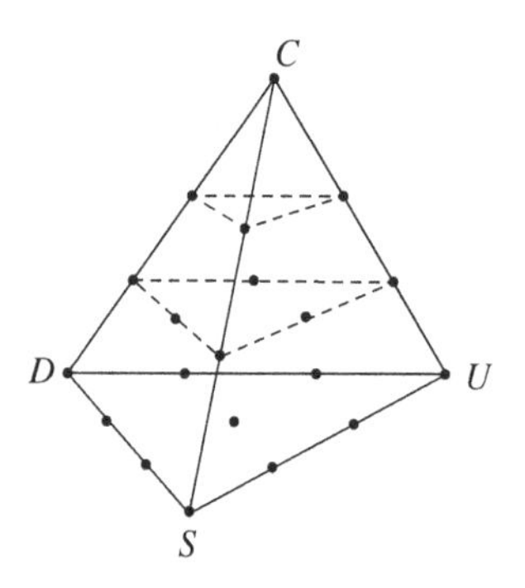

Figure 20-4. Interconnectivity for four flavors.

Accommodating the addition of another flavor (for a total of five) by trying to fit to another regular three-dimensional structure is ***precluded***, as discussed in [10], by the well-known **Euler** *constraint*,

$$\mathbf{v} + f - \mathbf{e} = 2, \tag{20-1}$$

where **v**, f and **e** stand for vertices, faces and edges as accentuated in the following summarization:

- Successive ***structures*** must be ***regular geometric*** entities that subsume the lower order structures.
- Each structure is determined by the ***combinatorics*** of the ***extrema*** (Within the Euler constraint for solids: $\mathbf{V} + \mathbf{F} - \mathbf{E} = \mathbf{2}$).

	v	f	e	Face shape
cube	8	6	12	square
octahedron	6	8	12	triangular
dodecahedron	20	12	30	pentagonal
icosahedron	12	20	30	triangular

However, the addition of ***two*** flavors, for a total of six, in other words, ***three*** *complete* generations, *can* be modeled by invoking the symmetries of the ***icosahedron*** whose vital statistics, known since antiquity, are (in consonance with Euler's equation) 12 vertices connected by 30 edges which form the boundaries of 20 equilateral triangular faces.

In Figure 20-5, we show, first for reference, a representative icoshedron, and then an example of a capital letter labeling scheme for (visible) vertices of an icosahedron spread out so that what appears to be a pentagonal outer perimeter is actually the representation of the single vertex labeled D^*. Note that there are indeed 20 faces represented, five whose corners bear fermion labels, five with antifermion labels, five with two fermion and one antifermion label and another five with one fermion and two antifermion labels. Of course each label is associated with five other vertices and an associated five other labels so that there are indeed $2 \times 6 = 12$ unambiguously labeled vertices and $2[6!/(2!4!)] = 30$ unambiguously labeled edges. In other words, all icosahedral ***vertices***, ***edges*** *and* ***faces*** are accounted for as are, correspondingly, all six fermionic ***labels*** and associated antifermionic ***labels***; a perfect fit!

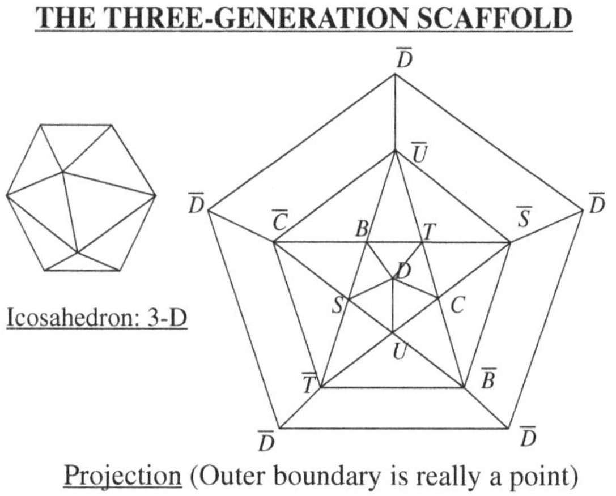

THE THREE-GENERATION SCAFFOLD

Icosahedron: 3-D

Projection (Outer boundary is really a point)

Figure 20-5. Reference and labeled icosahedron.

Furthermore, as noted in [11] "this *is* ***as far as we can go*** in label three space with the high degree of symmetry implied by the combinatorial requirements of a family structure". The implication is that there can be ***three generations*** but ***no more*** on the particle family tree.

There is a lot more to the "munificent geometry" of the icosahedron although not all of it is germane to this paper. However, it *is* pertinent to point out that vertices, edges and faces each come in ***antipodal*** pairs. Thus, looking directly down upon the vertex labeled D in Figure 20-5, one sees a pentagonal outline for the assembly of 5 faces formed by the connections between vertex D and each of the five surrounding vertices, labeled clockwise from the top as B, T, C, U and S. Also visible is the antipodal pentagonal outline formed by the connections between the antipodal vertex D^* and each of the associated antipodal vertices, labeled clockwise from the bottom as B^*, T^*, C^*, U^*, and S^*.

Note that the two pentagonal outlines (and their included faces) are in relative rotation about the DD^* axis by 36° which is half the central angle of an edge as measured in the pentagonal plane. Antipodal ***edges*** can then be related in ***pairs*** to form rectangles whose shape is such that the length, h, of the long side and the length, ℓ, of the short (edge) side are (it can be readily shown) in the "Golden ratio" $(1+\sqrt{5})/2$ and diagonals UU^* and DD^* have lengths of $\sqrt{5(1+\sqrt{5})/2}$ times the edge length. An example is the rectangle with edge UD connected to antipodal edge D^*U^* as in Figure 20-6. Following **Baez** [41] we shall henceforth refer to the rectangles as "***Duads***". Since there are 30 edges, there are 15 such duads.

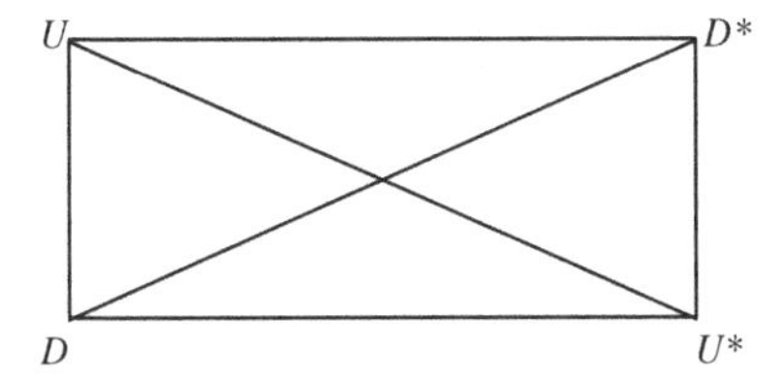

Figure 20-6. Example of a duad.

Also as per [41] the duads can be assembled into groups of three called "***Synthemes***" and the Synthemes can be assembled into three larger groups of five Synthemes each called "***Pentads***". One such pentad is the assembly of "*true crosses*", listed below, so called, in [41] because each of its five synthemes consists of three, *mutually orthogonal* duadic planes. In this list, each duad is represented by the labels of one of the edges with the opposite edge to be understood (for example, UD represents the rectangle shown in Figure 20-6, above). Note that there are 15 such duads in the list, each of which occurs once and each label occurs five times, that is, paired in a duad with each of the other five labels.

$$\begin{array}{ccc} UD & CS & TB \\ US & CB & TD \\ UB & CT & DS \\ UC & TS & DB \\ UT & CD & BS. \end{array}$$

Of particular interest because of its correspondence with the three generations of the SM is the ***first syntheme*** which is illustrated in Figure 20-7a. The three planes are indeed mutually orthogonal but also encompass all 12 icosahedral vertices. Thus, they completely *define* the icoshedron and its labeling. Alternatively, the icoshedron may be defined by the associated orthogonal *coordinate system* shown in Figure 20-7b where coordinate axes $\widehat{i}, \widehat{j}$ and $\widehat{k}$ are the ***normals*** to duads *UD*, *TB* and *CS*, respectively.

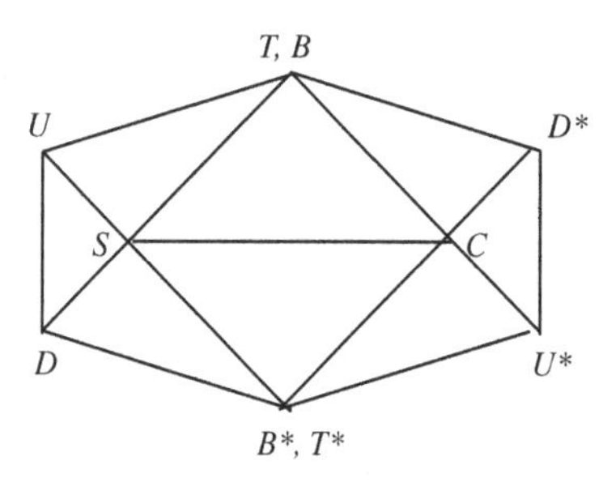

Figure 20-7a. First syntheme.

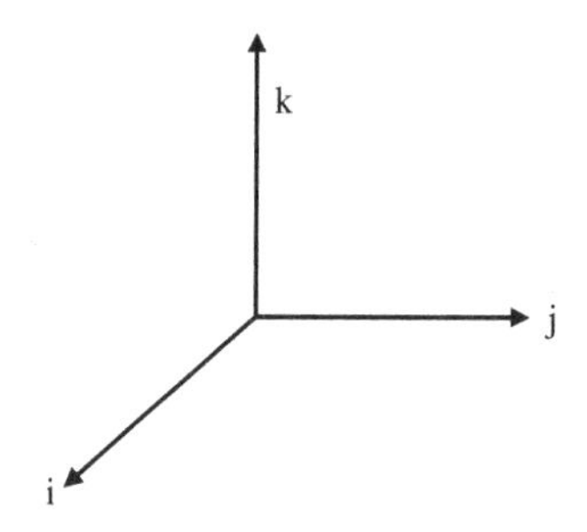

Figure 20-7b. Coordinate system for syntheme.

To round out this section, we also need to illustrate what it is that is emplaced upon loci in the triangular faces of the scaffolds discussed above. As mentioned, a prototypical triangular face can accommodate up to a ***decuplet*** of detailed alternate model particle models. We shall present an example of such a decuplet shortly but first, with reference to the DUS triangle of Figure 20-3, suppose we *ignore* the three *corner* locations of the face as shown in Figure 20-8. Although we are nominally left with the intermediate *seven*, in the process of ***descending*** from the top row to the second row by *selecting* a d quirk for replacement by an s quirk, the detailed ***symmetry*** of the alternate model is ***broken***, in particular that of

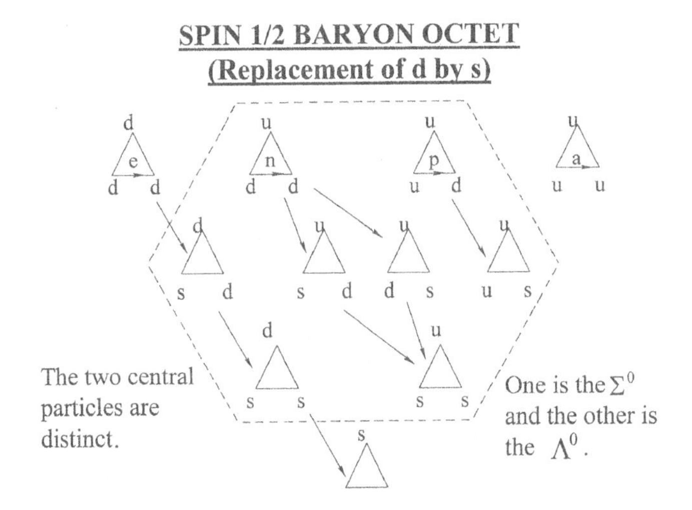

Figure 20-8. Spin 1/2 baryon octet.

the fermion labeled "n" for neutron. This allows the central location of the second row to harbor ***two distinct*** particles with the overall result that we realize the spin 1/2 baryon ***octet*** of the Standard Model. The two central particles are the alternate model's version of the $\sum^0$ and Λ^0 particles.

We can obtain a "mirror image" situation to this octet if we go back to the two contiguous central triangular faces shown in the labeling arrangement of Figure 20-5. Focusing on the *DUC* triangle instead of the *DUS* triangle (Figure 20-9) we can then visualize moving on a slant to the right from the *DU* locus as we replace u quirks by c quirks (rather than d quirks by s quirks). Thus we encounter the same *symmetry breaking*, but this time of the fermion labeled "p" on that locus. The two particles that result in the central location are then the alternate model version of the Standard Model's $\sum_c^+$ and Λ_c^+ particles.

Returning to the *decuplet*, Figure 20-10 shows the *spin 3/2 baryons* beginning with the delta particles, as portrayed in the top row. Only three quirks actually take part in the replacement process, so the *nonparticipating* quirks are shown enclosed by circles. We no longer have the triangular symmetry featured by the neutron

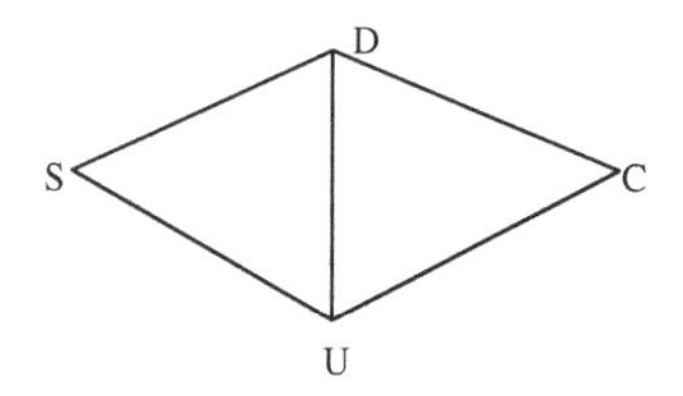

Figure 20-9. Mirror image triangle.

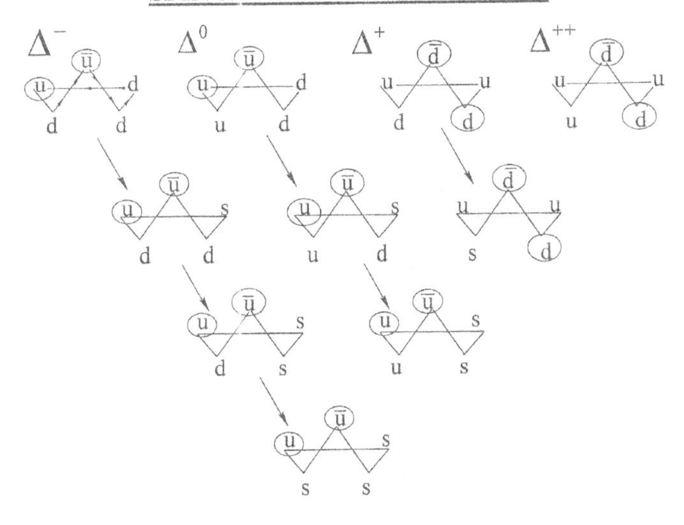

Figure 20-10. Spin 3/2 baryon decouplet.

in the octet of Figure 20-9 so *all transitions* in a given line are distinguishable. Thus, except for the three corners, all labeled loci are degenerate so that the figure really represents a multiplicity of diagrams, which, as pointed out in [10], mirrors SM practice for this decuplet.

As a final example, we show in Figure 20-11 another ***octet*** format, this time the *spin 0* ***meson*** octet of the SM, beginning the replacement process with the *neutral pion pair* as portrayed in the alternative model. Nonparticipating quirks are again indicated by circles but this time only a single quirk–antiquirk pair remains in

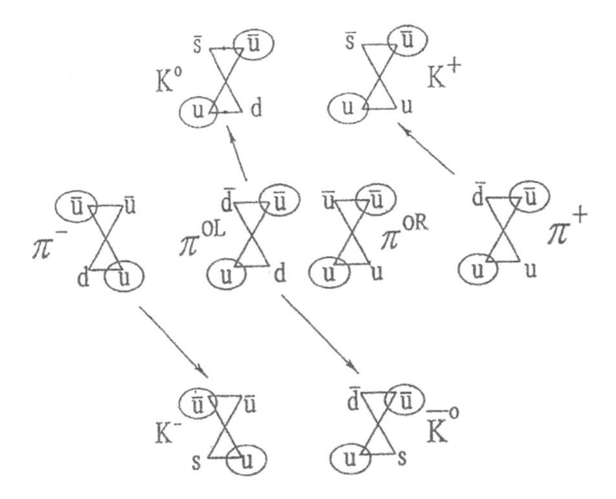

Figure 20-11. Meson octet.

each meson model to take part in the replacement procedure. Also, this time the procedure moves upward replacing d* antiquirks by s* antiquirks as well as downward replacing d quirks by s quirks as before.

V
Basic Concepts: Differential Geometry

Preamble: William Kingdon Clifford (4 May 1845–3 March 1879)

After I wrote most of this section it occurred to me that it would be fitting to say something here about Clifford beyond the mention in the Introduction; his views on space and matter were so strikingly prescient of what this book is about. Of course, we encountered his Clifford algebra in Sec. II, Chap. 4, but what is more germane here is where in the Introduction we talked about his views on the geometry of space. You may recall how he described matter as "little hillocks" of geometrical distortion that moved within and of it — very much like the basic ontological philosophy of this book and especially of this section; can you not just see those little hillocks wandering around?

Anyway, to find out more about him I went to the Internet and here is basically what I found out: Apparently Clifford was very much taken with the geometrical innovations of Riemann and Lobachevsky; they influenced his views in a fundamental way and in turn his views were a major influence on others, among whom we note, in particular, Professor John Wheeler whom you will meet again at the end of this section. In February 21, 1870, Clifford read a paper before the Cambridge Philosophy Society in which he discussed Remannian geometry to begin with but then he continued as follows:

"I wish to indicate the manner in which these speculations may be applied to the investigation of physical phenomena. I hold in fact

(1) That small portions of space are in fact of a nature analogous to little hills on a surface which is on the average flat; namely that ordinary laws of geometry are not valid in them.
(2) That this property of being curved or distorted is continually being passed on from one portion of space to another after the manner of a wave.
(3) That this variation of the curvature of space is what is really happening in that phenomenon we call the motion of matter, whether ponderable or ethereal.
(4) That, in the physical world, nothing else takes place but this variation, subject (possibly) to the laws of continuity."

I think that is amazing, do you not? I mean, 144 years ago! And 45 years before Einstein (who, as you will see along with Wheeler, also took note of Clifford's ideas).

21

Preliminaries

So, with that to cheer us on let us proceed with the work of the book. The previous sections were primarily concerned with the ***algebraic*** geometry of flattened MS and their interactions, that is, with FMS as *particles* but at this point, we embark upon a different kind of investigation; our particles are presumed to have a life before they become "particles", so, in this section we will be talking about the nature of these nascent particles which means mainly in terms of their ***differential*** geometry in one form or another. In the preceding, there was some fine-sounding talk about Toroidal Topology and about ***Solitons*** being, as it were "formed out of the dust of spacetime" (that allusion suddenly emerged unbidden and I could not resist it!).

But, seriously: how and why are our particles born and why do they persist as particles? And from a more comprehensive point of view, how and why do only those particular particle varieties emerge and why are all members of each variety exactly the same? Fair questions, but not necessarily confined to our model; those are the kinds of questions that have bedeviled scientists and philosophers for as long as there were such. How much light our alternative model can shed on the subject remains to be seen at this point in the book. (We will get back to it.) But as far as persistence is concerned, we know that solitons embody that quality — that is what solitons are supposed to do, and in fact we have some evidence of that for the basic particles of our concern here, the ones discussed in the early sections of the book.

By the way, at least in my opinion, Clifford's Assertion #2 in the Preamble on the preceding page is about as neat and concise a definition of a Soliton as one might want.

In any case, one of the main things we want to do in this section is to *demonstrate* the solitonic nature of our nascent particles and, moreover, to do it in a way that leads to some interesting physics. We begin with a simple model that lends itself to some analysis. You know, a mathematical line in three dimensions can curve around in any manner and not intersect itself. A line constrained to lie on a surface may or may not intersect itself and may or may not form a knot depending on the ***topology*** underlying the surface and any ***constraints*** it is subject to. There is also the possibility of the progression of a point along the surface in an ***unconstrained*** manner to trace out a so-called "***geodesic***", on the surface, a subject we shall address presently. In our case, being concerned with an MS prior to flattening, modeled as a concatenation of (2, n) torus knots, we begin, for convenience, with the admittedly ***idealized*** notion of an individual knot, actually wound around an actual torus as pictured in Sec. II and repeated here for reference. Again, I repeat: although it is most important to take a good look at a definite situation like this, there really ***is no torus*** to wrap around, down there in the micro–micro world. Nevertheless, we shall proceed as if there were, at least for a while!

So: a point on the toroidal surface is locatable by the vector we see in Figure 21-1, namely:

$$\mathbf{S} = \hat{i}x + \hat{j}y + \hat{k}z, \tag{21-1}$$

where

$$\begin{aligned} x &= w\cos\phi, \\ y &= w\sin\phi, \\ z &= r\sin\theta, \\ w &= R + r\cos\theta, \end{aligned}$$

ϕ is measured in the toroidal core's longitudinal (long way) direction and θ in its meridianal (short way) direction.

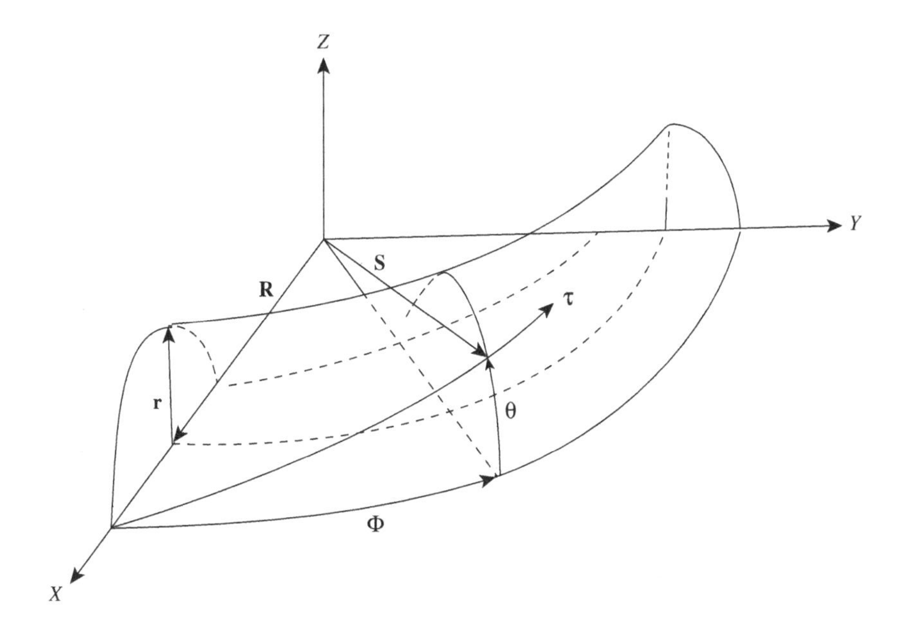

Figure 21-1. A section of a torus knot.

Going back to topology for a moment, in terms of wrapping a line around a surface several simple but definitive surface topologies come to mind namely the ***cylinder***, the ***sphere*** and the ***torus*** and, unfortunately, we are stuck with much the most complicated choice! Where the cylinder is curved in only one direction — its so-called *Gaussian* curvature [42] is zero (it can be cut and unrolled into a planar sheet) — the sphere has constant ***elliptic*** (positive) curvature. However, the torus is appreciably more complex than either because its topology is inherently a function of location; it exhibits ***elliptic*** curvature at the outer central latitudes ($w > R$), ***hyperbolic*** (negative) curvature at the inner central latitudes ($w < R$). In between, at $w \simeq R$, the poles of the sphere have become, for the torus, the two circles of polar latitude. The net result is that the differential geometric variables cannot be separated from the surface and wrapping parameters.

Of course, we are not complaining; all the neat toroidal things we talked about in the preceding sections far outweigh the problems! And here is an interesting sidelight, a unifying theme that runs through each of the wrapping situations mentioned above: central to each description are three parameters, two that define the *surface* and a generalized "***pitch***" *constraint* that defines the *wrapping.* Furthermore, note that another way to present the information in Eq. (21-1) is to write the location vector for the point on the torus as

$$\mathbf{S} = \mathbf{R} + \mathbf{r}, \tag{21-2}$$

where

$$\mathbf{R} = R(\hat{\imath}\ \cos\phi + \hat{\jmath}\ \sin\phi),$$
$$\mathbf{r} = r(\hat{\imath}\ \cos\theta\ \cos\phi + \hat{\jmath}\ \cos\theta\ \sin\phi + \hat{k}\ \sin\theta),$$

$\mathbf{r}$ is the locus of points on an equivalent sphere with radius r and $\mathbf{R}$ is the circular locus of the sphere's center. However, as we will see shortly, $\mathbf{R}$ can also be viewed as the projection onto base plane coordinates of a point on a cylinder of radius R. In a sense, the torus may thus be viewed as a surface *intermediate* between a cylinder and a sphere. As R goes to zero we have a sphere and as it goes toward infinity with finite r, a finite section of the toroidal core begins to look like a cylinder and the wrapping begins to look more like a circular helix.

The wrapping geometry of the cylinder is particularly simple and provides some insight into our main concern at this point, the wrapping of the torus. A simple, fixed constraint (see below) creates a helix, a portion of which is shown in Figure 21-2 wrapping in the ψ direction around a cylinder of radius r as it propagates up in z parallel to the $\hat{k}$ (cylinder) axis. Note that incremental ϕ for the case of the torus has become incremental z in this reparametrization (i.e. $R\delta\phi \rightarrow \delta z$). In what follows, we shall take a straightforward, elementary differential geometric approach wherein to each point on a curved line there is associated a unique ***frame*** [42], a coordinate system within which the attributes of curvature and torsion are defined. The figure shows a unit vector $\hat{t}$ ***tangent*** to the curve and $\hat{p}$ the principal ***normal*** to it. A ***binormal*** vector $\hat{b} = \hat{t} \times \hat{p}$ then completes the frame that moves along with the point on the curve whose locus is

$$\mathbf{S} = \hat{\imath}x + \hat{\jmath}y + \hat{k}z, \tag{21-3}$$

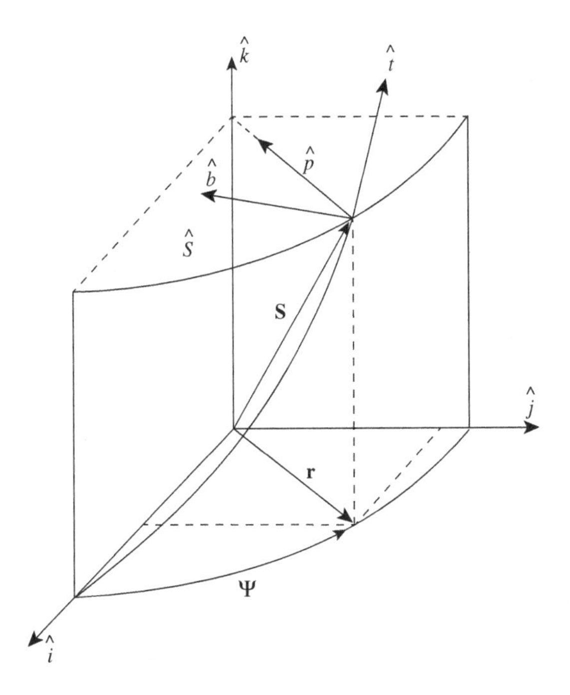

Figure 21-2. "Frame" coordinates for helix.

where

$$\begin{aligned} x &= r\cos\psi, \\ y &= r\sin\psi, \\ z &= c\psi, \\ c &= \frac{dz}{d\psi} = \frac{z_0}{2\pi} \end{aligned}$$

and z_0 is the "***pitch***" of the helix, the progression in the z direction during one full turn in ψ.

In more detail we find that

$$\begin{aligned} \hat{t} &\equiv \frac{d\mathbf{S}}{ds} \\ &= \frac{(-\hat{\imath}r\sin\psi + \hat{\jmath}r\cos\psi + \hat{k}c)}{(r^2+c^2)}, \end{aligned} \tag{21-4}$$

$$\hat{p} = -(\hat{\imath}\cos\psi + \hat{\jmath}\sin\psi), \tag{21-5}$$

$$\hat{b} = (\hat{\imath}c\sin\psi - \hat{\jmath}c\cos\psi + \hat{k}r) \tag{21-6}$$

and the line element associated with the helix is

$$ds = d\psi\sqrt{r^2 + c^2}. \tag{21-7}$$

Vector geometry is shown in Figure 21-3. Note that the principal normal is always perpendicular to the cylinder axis and in this figure is directed toward $-\hat{\imath}$. Angle γ (not drawn to scale) is given by $\gamma = \cot^{-1}\eta$ where $\eta = c/r$ is *pitch* normalized to the cylinder's circumference $2\pi r$.

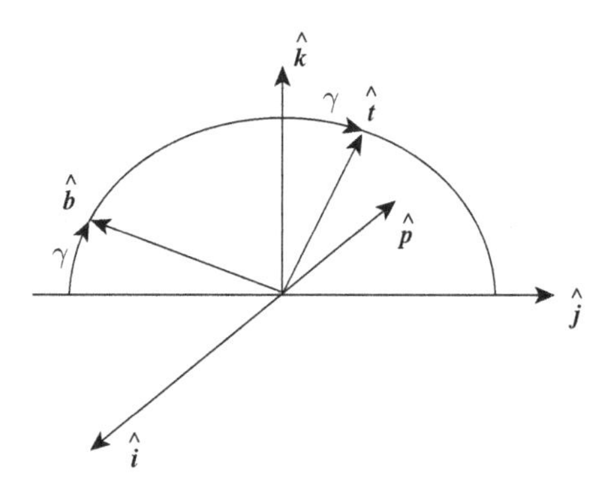

Figure 21-3. Vector geometry for the helix.

Also of interest are the **Frenet** relationships [98]

$$\begin{aligned} \frac{d\hat{t}}{ds} &= \kappa\hat{p}, \\ \frac{d\hat{k}}{ds} &= -\tau\hat{p}, \\ \frac{d\hat{p}}{ds} &= -\kappa\hat{t} + \tau\hat{b}. \end{aligned} \tag{21-8}$$

Here κ and τ are the "***curvature***" and "***torsion***" which, for the helix, are found to be

$$\begin{aligned} \kappa &= \frac{r}{(r^2 + c^2)} = \frac{1}{r(1 + \eta^2)}, \\ \tau &= \frac{c}{(r^2 + c^2)} = \frac{\eta}{r}(1 + \eta^2) = \eta\kappa. \end{aligned} \tag{21-9}$$

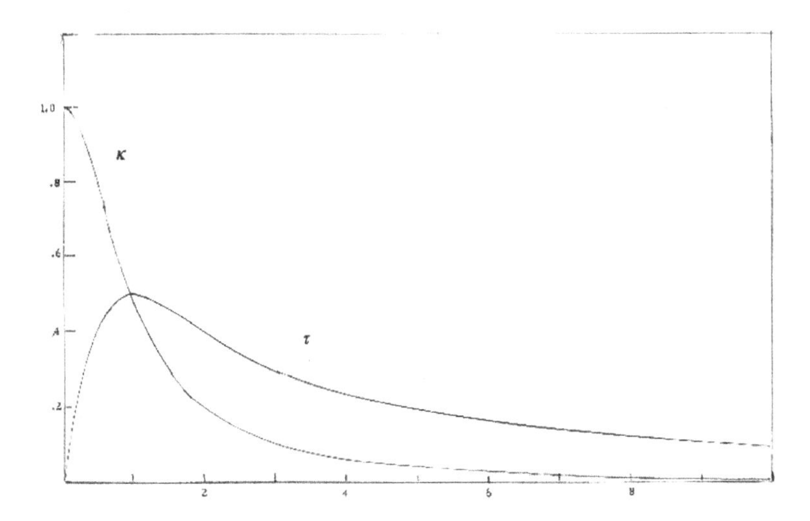

Figure 21-4. Curvature and torsion vs. pitch for helix.

Figure 21-4 shows curvature and torsion versus pitch for r fixed at unity. We see torsion increasing linearly with increasing pitch to begin with, peaking at $\eta = 1$ and eventually decaying as $1/\eta$ as the helix tends toward a straight line condition. Meanwhile, curvature is a maximum at $\eta = 0$ and decays steadily with increasing pitch as is to be expected. Although these parameters are strictly mathematical constructs there is at least a heuristic parallel to the stretching of a spring or to the notion of a particle being forced to move in the axial direction while it tries to circulate in the XY plane.

This point of view can be carried over to the differential geometry of the torus but, with an eye to where we are heading, it is convenient here to adopt a somewhat more general point of view, wherein the curved and twisted trajectory of a point moving along the knot evokes the notion of a ***connection*** to explicate the knot's topology. Connections are of course well-known as fundamental to the formulation of General Relativity. Actually, they are essential to most of modern particle theory, although the usual reference in physics is to "Gauge Theory" (Mainly due to the early influence of Professor Herman **Weyl**, but that is a long story [29]). In essence what's involved is the need to modify our familiar calculus in order to maintain the invariance of coordinate transformations of vectors that exist on non-Euclidean manifolds and this is accomplished by introducing additional space-dependent terms. Which, of course, is much too short and cryptic a description but we need to proceed here with our main story line. Fortunately, that includes an example: In GR, the terms on the left-hand side of the Einstein equation can be derived by the contraction of $R^{\mathrm{c}}_{\mathrm{dab}}$, the Riemann–Christoffel curvature tensor (RCCT), [43], a fundamental formalism we shall meet again, but which, in turn, can be seen to emerge from a consideration of the "parallel" transport of a vector between two points along each of the two different paths. The discrepancy that results between the two paths if they exist in non-Euclidean space is customarily formalized as the ***commutator*** of two covariant derivatives, each associated with its own path connecting the two points, viz:

$$(\nabla_a\nabla_b - \nabla_b\nabla_a)V^c = R^c_{dab}V^d + 2T^e_{ab}\nabla_e V^c, \qquad \text{(21-10)}$$

where V^{c} is the transported vector, the $\nabla_\alpha = \partial/\partial_\alpha + \Gamma^{\mathrm{e}}_{c\alpha}$ are the ***covariant*** derivatives and the $\Gamma^{\mathrm{e}}_{c\alpha}$ are the ***connections***, in particular, the ***Christoffel*** symbols (CS) of the second kind [42, 43]. While the first term on the R.H.S. is generally associated only with ***curvature***, the process of commutation also produces the additional quantity, $T^e_{ab} = (\Gamma^e_{ab} - \Gamma^e_{ba})/2$, the so-called Torsion Tensor. In conventional GR, this quantity vanishes due to the invariance of the CS to permutation of the first and second (lower) indices.

Historically, more general connections involving finite values of torsion have been considered, e.g. [44]. However, in what follows, finite torsion is seen to emerge automatically as a result of the explicit knot geometry. In any event, the task at hand is to compute the relevant *connections* and for that we recall the formalism used above, namely the location of a point on the toroidal surface by the vector we saw in the figure:

$$\mathbf{S} = \hat{i}x + \hat{j}y + \hat{k}z, \qquad \text{(21-11)}$$

where

$$x = w\cos\phi$$
$$y = w\sin\phi$$

$$z = r \sin \theta,$$
$$w = R + r \cos \theta,$$

ϕ is measured in the toroidal core's longitudinal (long way) direction and θ in its meridianal (short way) direction, $w = R + r\cos\theta$ is the projection in the x, y plane of the radius vector from the origin of coordinates to the point (ϕ, θ) on the toroidal surface, R is the radius of the toroid's circular centerline and r is the radius of its circular cross section.

For the general torus knot or link there is a definitive ***constraint*** between the two angles in terms of n, the number of circuits ***completed*** in the ***meridianal***, that is, the θ direction while ϕ ranges over m circuits in ***longitude***, defining what is then spoken of as an (m, n) torus knot or link. At this point, for (strictly) *pictorial* purposes, a convenient relationship, $\theta = n\phi/m$, will be assumed to apply over the extent of the knot. In the case of a $(2, n)$ knot (which we shall assume to prevail for the rest of the book), we thus have $\theta = n\phi/2$ so that

$$w = R + r\cos\left(\frac{n\phi}{2}\right), \quad z = r\sin\left(\frac{n\phi}{2}\right). \tag{21-12}$$

Note that for $\phi = 0$ we have the reference condition $w = R + r$ and $z = 0$. However, at the end of one circuit (i.e. for $\varphi = 2\pi$), we still have $z = 0$ but $w = R + r\cos(n\pi)$ which is equal to $R \pm r$, depending on whether n is ***even or odd*** and, as in the foregoing, leads to fermions for n odd and to bosons for n even. Varying ϕ for a fixed value of r then generates a visualizable representation of the knot or link as in Figure 21-5 for the particular value of $m = 2$ and $n = 3$ which defines the ***trefoil*** knot. As we see, there are three crossover points at which either of these two-dimensional ***projections*** is fundamentally ***ambiguous*** (although taken together they can be used to hypothesize an unambiguous trajectory).

The ambiguity evokes an interesting connection to the physics of the early 20th century. It involves Albert **Einstein**, who in 1917 wrote a paper [45] that according to [46] "— contained an elegant reformulation of the **Bohr–Sommerfeld** *quantization*

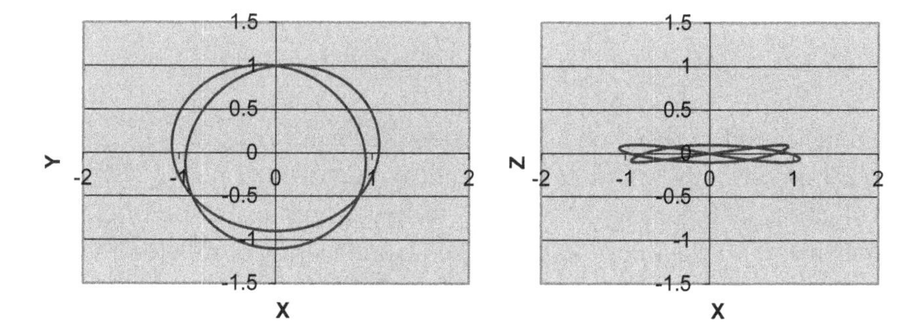

Figure 21-5. Planar trefoil projections.

rules of the old quantum theory —." At the time this was quite an important paper, cited by, among others, **De Broglie** in his thesis and **Schrödinger** in "the second of his seminal papers on the wave properties for matter". Although the paper's subject matter was considerably more general, of particular interest here is Einstein's observation that the self-intersecting momentum vector field generated by a particle moving under the influence of a central force field can be mapped onto the surface of a ***torus*** containing the the actual trajectory of the particle which, in our notation, is given by a particular history of the *planar* vector $\mathbf{w} = \hat{i}x + \hat{j}y$. As the particle circles around the XY plane (the ϕ progression) it oscillates in the radial direction between the two circles of radius $R - r$ and $R + r$ with a trajectory that intersects itself such that there is an inherent *ambiguity* in momentum; that is, at each crossover point, the momentum is double-valued.

However, in what might be termed "Einstein's ***ansatz***", incoming trajectories are mapped onto the *upper* surface of the torus and outgoing trajectories onto the *lower* half. Einstein shows that each half is quantized and that connecting the two then produces a trajectory that is single-valued. In terms of the torus knot description presented above, Einstein's mapping amounts to *adding* the vector $\mathbf{z} = \hat{k}\, r \sin\theta$ to the trajectory, so that the rate of change of $\mathbf{S} = \mathbf{w} + \mathbf{z}$ is *unambiguous*. In other words, as per our modeling Einstein's generalized trajectory for a particle in a central field of force takes the form of a ***torus knot*** for which it is only the *projection* onto a plane that is ambiguous as pictured in Figure 21-5 above. In what follows we will have occasion to invoke the ansatz in a very basic, but, one might say, somewhat radical way; very exciting, maybe even important!

Returning to the computation of the requisite connections, we begin with the CS of the *first* kind for a curve on a surface in three space defined as [42]

$$\Gamma_{abc} \equiv \left(\frac{\partial^2 \mathbf{S}}{\partial u^a \partial u^b}\right)\left(\frac{\partial \mathbf{S}}{\partial u^c}\right), \quad a,b,c = 1,2, \tag{21-13}$$

where, as above, a point on the curve is located by the vector $\mathbf{S}\{u^1(s), u^2(s)\}$, s being measured along the curve. Although there are nominally eight varieties of these symbols, two pairings, due to the symmetry of the second derivative, reduce this immediately to six, namely

$$\Gamma_{111}, \Gamma_{112}, \Gamma_{122}(=\Gamma_{212}), \Gamma_{121}(=\Gamma_{211}), \Gamma_{221} \quad \text{and} \quad \Gamma_{222}. \tag{21-14}$$

With reference again to the geometry of Figure 21-1 and associated location vector, setting $u^1 = \phi$ and $u^2 = \theta$, and carrying out the indicated derivatives (see Appendix A), we then find that all connection varieties vanish except for these two,

$$\begin{aligned} &1. \ \Gamma_{\phi\phi\theta} = rw\sin\theta, \\ &2. \ \Gamma_{\phi\theta\phi} = -rw\sin\theta. \end{aligned} \tag{21-15}$$

Note the *antisymmetrical* relationship between these two varieties as distinguished by the permutation of their *second* and *third* indices (*not* the first and second indices mentioned above whose invariance is cited as underlying the vanishing of torsion in conventional GR). The associated CS of the *second* kind is then

$$\Gamma^d_{ab} \equiv \Gamma_{abc} g^{cd}, \quad a,b,c,d = 1,2, \tag{21-16}$$

where the components of the metric tensor, g^{cd}, are

$$g^{11} = \frac{g_{22}}{g},$$
$$g^{22} = \frac{g_{11}}{g} \tag{21-17}$$
$$g^{12} = g^{21} = \frac{-g_{21}}{g}$$

g is the determinant $|g_{ab}|$ and $g_{ab} = (\partial \mathbf{S}/\partial u_{\mathrm{a}}) \cdot (\partial \mathbf{S}/\partial u_{\mathrm{b}})$. We then find that

$$g^{12} = g^{21} = 0,$$
$$g^{11} = \frac{1}{w^2},$$
$$g^{22} = \frac{1}{r^2}, \tag{21-18}$$
$$g = (rw)^2,$$

whereupon the CS of the second kind associated with Eq. (21-5) are

$$1.\ \Gamma^{\theta}_{\phi\phi} = \left(\frac{w}{r}\right) \sin\theta,$$
$$2.\ \Gamma^{\phi}_{\phi\theta} = -\left(\frac{r}{w}\right) \sin\theta. \tag{21-19}$$

Although there is no longer an antisymetrical (or symmetrical) relationship between the two CS varieties, to provide some perspective it is still of interest to form their difference

$$\Delta\Gamma \equiv \Gamma^{\theta}_{\phi\phi} - \Gamma^{\phi}_{\phi\theta} = \Delta\Gamma = \left(\frac{w^2 + r^2}{rw}\right) \sin\theta, \tag{21-20}$$

which, we can then relate to elementary differential geometric notions of curvature and torsion. To begin with we note that

$$(ds)^2 = w^2(d\phi)^2 + r^2(d\theta)^2. \tag{21-21}$$

So that

$$\frac{ds}{d\phi} = (w^2 + \mu^2 r^2)^{1/2} \tag{21-22}$$

is the rate of change of curve length with respect to traverse in longitude at latitude θ and $\mu = d\theta/d\phi$. Then, defining

$$\begin{aligned} w &\equiv \left(\frac{ds}{d\phi}\right)\cos\alpha, \\ r &\equiv \left(\frac{ds}{d\phi}\right)\sin\alpha, \end{aligned} \tag{21-23}$$

we have

$$\Delta\Gamma = \frac{\sin\theta\alpha}{\sin\alpha\cos\alpha}, \qquad \alpha = \tan^{-1}\left(\frac{\mu r}{w}\right). \tag{21-24}$$

The relationship between small increments $\mu r\Delta\phi$ along the meridian, $w\Delta\phi$ along the circle of latititude and the curve itself, Δs, due to a small increment in longitude, $\Delta\phi$, is shown in Figure 21-6.

Neglecting the $\sin\theta$ term for the moment, it is instructive to think of the ***local*** behavior in terms of an *equivalent* ***helix*** wrapped around a cylinder of radius r such that traverse in the direction of the cylinder's axis by an amount $w\Delta\phi$ is accompanied by a circumferential increment $\mu r\Delta\phi$. Angle α is thus the ***inclination*** of the coiled line to the cylinder's axis. As above, we found the curvature and torsion for the helix to be given in differential geometric terms as

$$\begin{aligned} \kappa &= \frac{r}{\left[\left(\frac{z_0}{2\pi}\right)^2 + r^2\right]}, \\ \tau &= \frac{\left(\frac{z_0}{2\pi}\right)}{\left[\left(\frac{z_0}{2\pi}\right)^2 + r^2\right]}, \end{aligned} \tag{21-25}$$

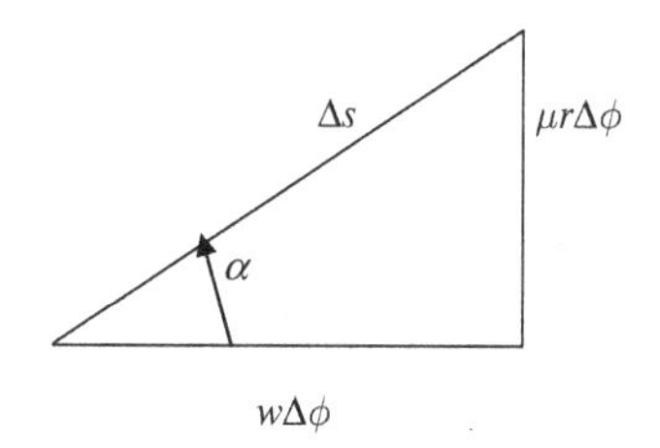

Figure 21-6.

where, we recall, z_0 is the "***pitch***", the amount of traverse of a point *along* the cylinder's axis during a full turn *around* it. In this picture, curvature is the magnitude of the rate of turn of $\hat{n}$, the unit vector normal to the helical curve (and to the cylinder axis), with respect to motion along the helical curve, while torsion is, similarly, the magnitude of the rate of change of the unit binormal $\hat{b} = \hat{n} \times \hat{t}$, where $\hat{t}$ is the unit vector tangent to the curve.

Now, for the ***torus***, suppose we make the association $\{z_0 \to 4\pi w/\mu\}$: we can then, similarly, associate

$$\begin{aligned} \kappa &\longrightarrow \frac{\mu^2 r}{w^2 + \mu^2 r^2} \\ \tau &\longrightarrow \frac{\mu w}{w^2 + \mu^2 r^2}. \end{aligned} \tag{21-26}$$

Thus, for *torus knots* in this picture, the ratio of curvature to torsion is $\kappa/\tau \to \mu r/w$, an important parameter. Also, as for the helix encircling its cylindrical core, a simple (but by no means trivial) way to view what's happening is in terms of a miniscule, hypothetical point entity trying to encircle the toroidal core along a meridian while constrained to travel in the longitudinal direction as well. However, where for the ***helix***, encirclement around the core is associated with ***curvature*** and the displacement due to continual motion in the direction of the helical axis is the source of ***torsion***, for the ***knot***, although quite similar, the situation is a little more involved as we shall see due to the additional toroidal topology.

Finally, consider again the antisymmetrical relationship of the two nonzero CS of the first kind, which, as noted above, is associated with permutation of the *second and third* CS indices. This suggests defining an "*alternative*" Contortion Tensor as

$$K_a \equiv \frac{\Gamma_{\phi\phi\theta} - \Gamma_{\phi\theta\phi}}{2} \tag{21-27}$$

and on that basis, the $\Delta\Gamma$ of Eq. (21-15) can be viewed as

$$\Delta\Gamma = 2(g^{\phi\phi} + g^{\theta\theta})K_a, \tag{21-28}$$

which, according to the above is also equal to

$$\Delta\Gamma \to \left(\frac{1 + \left(\frac{\kappa}{\tau}\right)^2}{\frac{\kappa}{\tau}}\right) \sin\theta. \tag{21-29}$$

Also, for some insight into the results of the next chapter, we note that

$$\Gamma^{\theta}_{\phi\phi} = g^{\theta\theta} K_a. \tag{21-30}$$

22

Solitonic Behavior; The Geodetic Approach

Although a torus knot would appear to be topologically ineradicable, it is of interest to explore its *explicit* behavior from the point of view of its description, as per the Introduction, as a *disturbance* in the continuum. We begin with the notion of a miniscule, representative disturbance, moving *along the knot* according to the canonical equation for a ***geodesic*** [43],

$$\frac{d^2x^a}{ds^2} + \Gamma^a_{bc}\left(\frac{dx^b}{ds}\right)\left(\frac{dx^c}{ds}\right) = 0, \tag{22-1}$$

even though, we note, there is no prior reason to assume that an arbitrary knot constrained to lie on the surface of a real torus will obey that relationship. Then, using the description of our knot, plus the definitions

$$\begin{aligned} x^a &= \theta, \\ x^b &= x^c = \phi, \\ \Gamma^a_{bc} &= \Gamma^\theta_{\phi\phi} = \left(\frac{w}{r}\right)\sin\theta \text{ (as per Eq. (22-19))}, \\ ds &= (w^2 + \mu^2 r^2)^{1/2} d\phi, \text{ the line element}, \end{aligned} \tag{22-2}$$

this becomes

$$\frac{d^2\theta}{ds^2} + \eta^2 \sin\theta = 0, \tag{22-3}$$

where

$$\eta = \left[\frac{w}{r}(w^2 + \mu^2 r^2)\right]^{1/2}$$

which is $\eta \cong 1/\sqrt{wr} \approx 1/\sqrt{Rr}$ to first order in r/R. Note that, on the basis of Eqs. (22-8) and (22-9), we can also write $\eta \cong 1/g$ and henceforth designate the length $g = \sqrt{rR}$ as "*The Metric*".

A caveat: Equation (22-3) is not *exactly* the *geodetic* equation for a point encircling a torus but it is a very good approximation to it for *slender* tori (see below) and, as we shall see presently, the tori with which we are concerned are, indeed, very slender. In any event, we recognize it as the ***sine–Gordon*** *equation* whose solution is well known, and readily verifiable (read on), as describing Solitonic behavior in the form

$$\theta = \pi - 4\,\arctan(e^{-\eta s}), \tag{22-4}$$

indicating that θ varies symmetrically about $\theta = 0$, from $-\pi$ to π, as s varies from $-\infty$ to ∞ with most of the variation taking place in a length increment $\Delta l = 1/\eta = \sqrt{g}$. As a fraction of the length of an actual MS, this increment is $\Delta l/2\pi R$, which, for small r/R is $\approx \sqrt{r/R}/2\pi$ and is expected to be quite small. The variation thus has the shape of a sharply rising S-curve, the rate of rise depending on the ratio, R/r which according to the above is identifiable as the ratio of torsion to curvature for the helical equivalent of the torus. Note also that twist information as manifested by the factor μ has been suppressed in going to the small r/R condition implying that sine–Gordon behavior is the same for all orders of twist.

Actually, the subject of geodesics on a torus is somewhat complex but, beginning with the connections as presented in the preceding, a well-known formula can be derived that leads to families of geodesics. For example, beginning with a formalism for location on the torus and the consequent connection terms identical to those we generated in the preceding, Irons [47] expresses the referenced formula in terms of the following pair of equations (his notation):

$$\ddot{\nu} + \left(\frac{1}{a}\right)\dot{u}^2(c + a\cos\nu)\sin\nu = 0,$$

$$\ddot{u} - \left(\frac{2a\sin\nu}{c + a\cos\nu}\right)\dot{u}\dot{\nu} = 0. \tag{22-5}$$

Then, if we make the identifications

$$\begin{aligned} \nu &\leftrightarrow \theta \\ u &\leftrightarrow \phi \\ t &\leftrightarrow s \\ c &\leftrightarrow R \\ a &\leftrightarrow \mu r \end{aligned} \tag{22-6}$$

we get

$$\frac{d^2\theta}{ds^2} + \left(\frac{w}{\mu r}\right)\left(\frac{d\phi}{ds}\right)^2 \sin\theta = 0 \tag{22-7}$$

and

$$\frac{d^2\phi}{ds^2} - \left(\frac{2\mu r \sin\theta}{R + \mu r \cos\theta}\right)\dot{\theta}\dot{\phi} = 0, \tag{22-8}$$

respectively, for the above two. Note that the first equation would be in standard sine–Gordon form if we could do something about the $(d\phi/ds)^2$ term on the R.H.S. There is a short way and a long way to do that. Just for fun we can outline the long way as follows: multiply the second equation by $d\phi/ds$ and then solve the first equation for $(d\phi/ds)^2 \sin\theta$ to get (after a short finagle)

$$\begin{aligned} \frac{d\left(\frac{d\phi}{ds}\right)^2}{ds} &= -\left(\frac{r}{w}\right)^2\left(\frac{d^2\theta}{ds^2}\right)\left(\frac{d\theta}{ds}\right) \\ &= -\left(\frac{r}{w}\right)^2 \frac{d\left(\frac{d\theta}{ds}\right)^2}{ds}. \end{aligned} \tag{22-9}$$

Integration then leads to (what might be called) the "canonical" expression for the line element except for a constant of integration, k, that multiplies the L.H.S.

$$k(ds)^2 = w^2(d\phi)^2 + r^2(d\theta)^2. \tag{22-10}$$

However, we must have $k = 1$ because the equation is true that way, so what we have accomplished is just to realize the *short way* mentioned above, which, for small r/w just says that

$(d\phi/ds)^2 \approx 1/w^2$ and leads, via substitution, back into the first equation above to yield

$$\frac{d^2\theta}{ds^2} + \left(\frac{1}{wr}\right) \sin\theta = 0, \tag{22-11}$$

our original Eq. (22-3); all very gratifying. In the meantime, however, it is important for what follows to verify sine–Gordon behavior another way, namely via a Lagrangian Approach.

23

The Lagrangian Approach

Equation (22-3) can also be derived from a Lagrangian but before we do so it is useful to talk about the sine–Gordon equation briefly. It appears in a variety of contexts but the canonical situation is the time-honored case of an idealized pendulum consisting of a stiff, weightless rod of length l and a weight of mass m constrained to rotate in a plane as portrayed in Figure 23-1 under the influence of gravity.

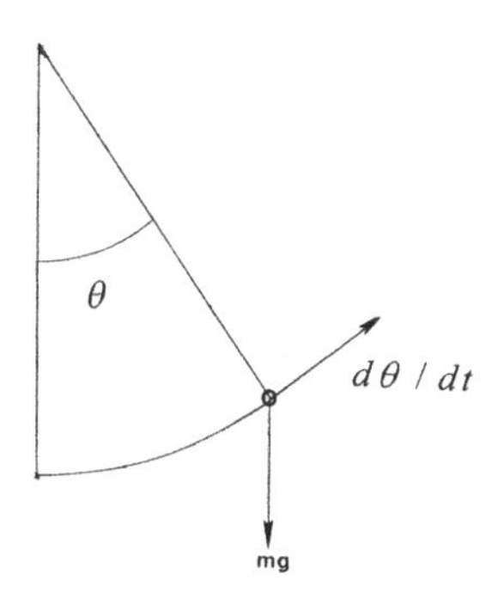

Figure 23-1. Pendulum.

The behavior of this elementary device is readily derived via straightforward Newtonian ($F = ma$) mechanics but specialized here, to angular coordinates where the restoring torque, $-(mgl)\sin\theta$, equals the moment of inertia, ml^2, multiplied by the angular acceleration, $d^2\theta/dt^2$, i.e.

$$ml^2\left(\frac{d^2\theta}{dt^2}\right) = -(mgl)\sin\theta, \tag{23-1}$$

or, in sine–Gordon format,

$$\left(\frac{d^2\theta}{dt^2}\right) + \left(\frac{g}{l}\right) \sin\theta = 0, \tag{23-2}$$

which can also be derived from a Lagrangian in a relatively straightforward way. The lagrangian is $L = K - V$, where K is kinetic energy, $K = (1/2)ml^2\dot{\theta}^2$, which, obviously, depends only on the rate of change of θ and V is potential energy, $V = mg(l - l\cos\theta)$ which depends only on θ itself. The corresponding Euler–Lagrange equation [37],

$$\frac{d\left(\frac{\partial L}{\partial\dot{\theta}}\right)}{dt} - \frac{\partial L}{\partial\theta} = 0, \tag{23-3}$$

then becomes

$$\frac{d\left(\frac{\partial K}{\partial\dot{\theta}}\right)}{dt} + \frac{\partial V}{\partial\theta} = 0. \tag{23-4}$$

That is

$$ml^2\left(\frac{d^2\theta}{dt^2}\right) + (mgl)\sin\theta = 0, \tag{23-5}$$

in other words, the same as what we got via the Newtonian approach. The proper Lagrangian to begin with is therefore

$$L = \left(\frac{1}{2}\right) ml^2 \left(\frac{d\theta}{dt}\right)^2 - mg(l - l\cos\theta) \tag{23-6}$$

or, since the constant term plays no part

$$L = \left(\frac{1}{2}\right) ml^2 \left(\frac{d\theta}{dt}\right)^2 + mgl\cos\theta. \tag{23-7}$$

Thus, the message for constructing a Lagrangian for the ***knot*** situation is that we need to start with something like

$$L = \left(\frac{1}{2}\right) A \left(\frac{d\theta}{ds}\right)^2 + B\cos\theta, \tag{23-8}$$

where, in this case, the derivative is with respect to knot length parameter s rather than time t, A has the dimensions of *energy per unit length* and, correspondingly, B is *energy per unit volume.* Both parameters are characteristic of the local spacetime but, are as yet unspecified.

That being the case, let us pause and consider: since we believe our particles are solitons, that is, deformations "in-and-of spacetime" as characterized in [10] and "continuously being passed on from one portion of space to another after the manner of a wave" as per Clifford, we should be invoking General Relativistic relationships. Actually, as far as A is concerned, we have that immediately according to

$$A = \frac{c^4}{4\pi G}, \tag{23-9}$$

where c is the velocity of light and G is the gravitational constant [20], ***nominally*** (but see Chap. 26) equal to $6.673 \times 10^{-11}\,\mathrm{m}^3/\mathrm{kg} \cdot \mathrm{s}^2$.

To specify B we embark on a whirlwind tour of General Relativity, invoking the notion that contraction of the Einstein field equation,

$$R_{\mu\nu} - \frac{1}{2} g_{\mu\nu} R = -8\pi G T_{\mu\nu}, \tag{23-10}$$

leads to the following relationship between energy and curvature, namely (cf. [43]):

$$R = \left(\frac{8\pi G}{c^4}\right) T^{\mu}_{\mu} = \frac{2T^{\mu}_{\mu}}{A}. \tag{23-11}$$

Here $R = g^{\mu\kappa} R_{\mu\kappa}$ is the Curvature Scalar (CvS), where

$$R_{\mu\kappa} = R^{\lambda}_{\mu\lambda\kappa} = \left(\frac{\partial \Gamma^{\lambda}_{\mu k}}{\partial x^{\lambda}} - \frac{\partial \Gamma^{\lambda}_{\mu\lambda}}{\partial x^{k}}\right) + (\Gamma^{\eta}_{\mu k}\Gamma^{\eta}_{\lambda\eta} - \Gamma^{\eta}_{\mu\lambda}\Gamma^{\eta}_{k\eta}) \tag{23-12}$$

is the Ricci ***Curvature*** Tensor (RCT), contracted from the RCCT we mentioned above

$$R^{\lambda}_{\mu\nu\kappa} = \left(\frac{\partial \Gamma^{\lambda}_{\mu\nu}}{\partial x^{k}} - \frac{\partial \Gamma^{\lambda}_{\mu k}}{\partial x^{\nu}}\right) + (\Gamma^{\eta}_{\mu\nu}\Gamma^{\eta}_{k\eta} - \Gamma^{\eta}_{\mu k}\Gamma^{\eta}_{\nu\eta}) \tag{23-13}$$

and

$$T_{\mu}^{\mu} = T_{\mu\nu} g^{\mu\nu} \tag{23-14}$$

is the Energy Scalar, contracted from $T_{\mu\nu}$, the ***Energy–Momentum*** Tensor (EMT). Since, as per (22-10), only two of the metric terms are nonzero, we have (with superscripts $1 = \phi$ and $2 = \theta$)

$$R = g^{\phi\phi} R_{\phi\phi} + g^{\theta\theta} R_{\theta\theta}. \tag{23-15}$$

The two terms in this equation are found to be identical (see Appendix C) whereupon their sum, the CvS, becomes

$$R = \left(\frac{2}{wr}\right) \cos\theta, \tag{23-16}$$

whose denominator (in contrast to the helix) embodies ***both*** *toroidal radii*, as is to be expected. Also, we note that the magnitude of curvature (as per the cosine factor) is a maximum on the two toroidal equators and vanishes on the polar circles where curvature changes from one sign to another, also as to be expected.

Actually, what we are really interested in here is expressing the Energy Momentum Tensor in terms of the Curvature Scalar, the point of view being that, in this case, ***energy*** is due to ***curvature*** (rather than the usual other way around). Combining (23-10) and (23-16) we then have

$$T_{\mu}^{\mu} = \left(\frac{A}{wr}\right) \cos\theta, \ \mathrm{kg(ms)^2/m^3} \tag{23-17}$$

which we set equal to the $B\cos\theta$ in Eq. (23-8) to give us the Lagrangian in the form

$$L = A\left[\left(\frac{1}{2}\right)\left(\frac{d\theta}{ds}\right)^2 + \left(\frac{1}{wr}\right)\cos\theta\right]. \tag{23-18}$$

The corresponding Euler–Lagrange equation [37] is thus

$$\frac{d^2\theta}{ds^2} + \left(\frac{1}{wr}\right)\sin\theta = 0 \tag{23-19}$$

which, to first order in r/R (recall that $w = R + \mu r\cos\theta$), is Eq. (22-3), in the previous chapter.

24

Solitonic Antiparticles

We recall a fundamental attribute of particle physics, namely that to every elementary particle there is associated a conjugate antiparticle. This concept was, of course, the landmark contribution of Paul ***Dirac***, realized as a result of his mating of quantum theory with the generalized energetics implied by special relativity (see Sec. II). Although particles are *visualized* herein as solitonic disturbances "in and of an otherwise undistorted continuum", to begin with, there was no mention, to this point, of what it is that does the disturbing. Now we see that it is the traversal by the solitonic *torsional* ***distortion*** around the putative toroidal core that does so. However, note that there are ***two directions*** of traverse available in the model employed herein and, in effect, only one has actually been invoked, by implication, traverse to the ***right*** (as per Figure 21-1). Thus it seems natural to ask whether our *solitonic model* can also accommodate the notion of ***antiparticles*** by invoking traverse to the *left*. Although this is the nominal state of affairs in Sec. II, it was occasioned by the *requirements* of fusion. Here, we ask whether that requirement is satisfied inherently.

To pursue this possibility it is expedient to first demonstrate the validity of the cited solution to the sine–Gordon equation. First, we note that, in what follows, the dependent variable θ of the sine–Gordon equation and its first and second derivatives with the independent variable s will be designated by θ_R, θ'_R and θ''_R, respectively (with R explicitly signifying traverse to the right). We then begin the demonstration with the following modification of Eq. (22-4):

$$e^{-\eta s} = \tan\left(\frac{\pi - \theta_R}{4}\right). \tag{24-1}$$

Differentiating both sides with respect to s then produces

$$\begin{aligned} -\eta e^{-\eta s} &= \frac{\left(-\frac{\theta_R'}{4}\right)}{\cos^2}\left(\frac{\pi - \theta_R}{4}\right) \\ &= 2\eta \sin\left(\frac{\pi - \theta_R}{2}\right). \end{aligned} \tag{24-2}$$

Solving for θ_R', (likewise) produces

$$\begin{aligned} \theta_R' &= 4\eta \sin\left(\frac{\pi - \theta_R}{4}\right)\cos\left(\frac{\pi - \theta_R}{4}\right) \\ &= 2\eta \sin\left(\frac{\pi - \theta_R}{2}\right) \end{aligned} \tag{24-3}$$

and another differentiation then yields the desired result:

$$\begin{aligned} \theta_R'' &= 2\eta \cos\left(\frac{\pi - \theta_R}{2}\right)\left(-\frac{\theta_R'}{2}\right) \\ &= -\eta^2 \sin(\pi - \theta_R) \\ &= -\eta^2 \sin\theta_R \end{aligned} \tag{24-4}$$

or, as required:

$$\theta_R'' + \eta^2 \sin\theta_R = 0. \tag{24-5}$$

For the case of traverse to the left, we therefore replace s by $-s$, and dependent variable θ with θ_L. We then begin as above with a corresponding replacement for Eq. (24-1), namely

$$e^{\eta s} = \tan\left(\frac{\pi - \theta_L}{4}\right). \tag{24-6}$$

Differentiating and solving for θ'_L produces

$$\theta'_L = -2\eta \sin\left(\frac{\pi - \theta_L}{2}\right) \tag{24-7}$$

instead of Eq. (24-3) whereupon, with a second differentiation and corresponding simplification we get

$$\theta''_L - \eta^2 \sin\theta_L = 0 \tag{24-8}$$

which is Eq. (24-5) with a minus sign instead of a plus sign between the two terms or, put another way,

$$-\theta''_L + \eta^2 \sin\theta_L = 0 \tag{24-9}$$

which we will use in the next chapter to complete our discussion of solitonic antiparticles.

Finally, it is instructive to demonstrate the extension of the solitonic behavior into the *relativistic* regime. As discussed above, most of the change in θ (and therefore most of the mass) takes place in a small region of s, namely $\Delta s = 1/\eta$. Thus, we consider the motion of that region over the range of s, which, for all practical purposes means along the longitudinal range of the torus. If the speed, v, of that motion is relativistic, a stationary observer will see a transformed velocity and an associated *modified* behavior given by

$$\theta = \pi - 4 \arctan\left\{e^{-\gamma\eta(s-vt)}\right\}, \tag{24-10}$$

where $\gamma = 1/\sqrt{1-(v/c)^2}$. Then, computing applicable derivatives of this behavior (as sketched above to verify the validity of the solution to the sine–Gordon equation) we find that this *modified* behavior is the solution to a *modified* sine–Gordon equation, namely

$$\left\{\frac{\partial^2\theta}{\partial s^2} - \left(\frac{1}{c^2}\right)\frac{\partial^2\theta}{\partial t^2}\right\} + \eta^2 \sin\theta = 0. \tag{24-11}$$

The expression in brackets will be recognized as the *Lorentz* derivative, a relativistic quantity, *invariant* to velocity changes.

25

Particle Parameters: Mass and Size

The objective here is twofold; first to derive some measurable quantities associated with our particle model but also, to show how a leftward traverse implies a negative mass. We begin by multiplying Eq. (24-5) by $Ar(v/c)^2(\Delta\ell)$ where $v = ds/dt$ is velocity along the knot. The result, again to first order in r/R, is

$$\left[\left(\frac{A}{c^2}\right)\sqrt{\mu Rr}\right](r\ddot{\theta}_R) + \left[\left(\frac{v}{c}\right)^2\sqrt{\frac{r}{R}}\right]\left(\frac{A}{2}\right)\sin\theta_R = 0, \qquad (25\text{-}1)$$

which is in the "Newtonian dynamic" form

$$m_R a_R + F_R = 0, \qquad (25\text{-}2)$$

where

$$\begin{aligned} m_R &= \left(\frac{A}{c^2}\right)\sqrt{Rr} = c^2 g/4\pi G\,\mathrm{kg} \\ a_R &= r\ddot{\theta}_R\frac{\mathrm{m}}{\mathrm{s}^2} \\ F_R &= \left(\frac{v}{c}\right)^2\sqrt{\frac{r}{R}}\left(\frac{A}{2}\right)\sin\theta_R\frac{\mathrm{J}}{\mathrm{m}}. \end{aligned}$$

Note that acceleration varies as $a_R = \left(\frac{v^2}{R}\right)\sin\theta_R$ and that mass is proportional to the ratio of the metric (as defined above) to the gravitational constant G (which is essentially the inverse of the linear energy density term A — see Chap. 26). On the other hand, multiplying Eq. (24-9) by that same expression, $Ar(v/c)^2(\Delta\ell)$, we get an equation analogous to the "dynamic equation" of Eq. (25-2), namely

$$-m_R a_L + F_L = 0, \qquad (25\text{-}3)$$

where, as per the above (but with modifications):

$$m_R = \left(\frac{A}{c^2}\right)\sqrt{Rr}\,\text{kg}.$$

$$a_L = r\ddot{\theta}_L\,\text{m/s}^2$$

$$F_L = \left(\frac{v}{c}\right)^2 \sqrt{\frac{r}{R}}\left(\frac{A}{2}\right)\sin\theta_L \frac{\text{J}}{\text{m}}.$$

That is, m_R is the same as m in Eq. (23-2) but both a_L and F_L incorporate θ_L rather than θ. Thus, in order to recreate the *format* of the dynamic equation, that is

$$m_L a_L + F_L = 0, \qquad (25\text{-}4)$$

we need to ***define*** the mass term for ***leftward*** traverse as $m_L = -m_R$.

Note that everything we have done so far refers to a ***particular***, *solitary torus knot* for which the radius of the implied toroidal core is r. But as noted in the Introduction, we are ultimately interested in particles in the form of Möbius strips which, visualized as ***concatenations*** of torus knots, suggests that the radius of the implicit toroidal core takes on a continuous range of values as in the pictured concatenation of torus knots in Sec. II. On this basis, expressing the mass of the kth knot as

$$m_k = \left(\frac{A}{c^2}\right)\sqrt{Rr_k} \qquad (25\text{-}5)$$

leads, on the basis of the preceding definition, to an expression for the mass of the MS as

$$m = \left(\frac{A}{c^2}\right)\sqrt{R\Delta r}\sum_0^N \sqrt{k}, \qquad (25\text{-}6)$$

where $N = W/\Delta r$ is the number of knots stacked in MS width W, Δr being the nominal spacing (yet unspecified) between knots. Then, approximating the summation by an integral we get

$$m \cong 2\left(\frac{A}{c^2}\right)\sqrt{R}\int_0^W \sqrt{r}\,\frac{dr}{\Delta r}$$

$$= \left\{\frac{4\left(\frac{A}{c^2}\right)W^{3/2}}{3\Delta r}\right\}\sqrt{R}.$$

$$= \left(\frac{4NA}{3c^2}\right)\sqrt{WR}. \tag{25-7}$$

The main point here is that, since MS mass varies in an explicit way with the radius of the toroidal core's centerline, we can compute the relative "***size***" of a pair of modeled "particles" if we *specify their* ***masses***, under the (as it turns out reasonable) assumption that $N = W/\Delta r$ and A are common to all particles. For example, we know the masses of the proton and electron quite well, the "size" of the proton not quite so well but the "size" of the electron not well at all. So, given proton and electron masses as

$$m_p = 1.673 \times 10^{-27}\,\text{kg} \quad \text{and} \quad m_e = 9.12 \times 10^{-31}\,\text{kg}$$

(for a ratio of about 1836), and a proton "size" of $R_p \cong 2 \times 10^{-15}$ m, we find the "size" of a modeled "electron" to be

$$R_e = \left(\frac{m_e}{m_p}\right)^2 R_p \simeq 2 \times 10^{-22}\,\text{m}$$

a value currently beyond the range of instrumentation but hopefully not forever, perhaps with the help of clever experimental design.

26

Mass, the Higgs and the Strong Force

The previous section considered particle mass only in a relative way in terms of its variation with particle size. Here we are concerned with the actual ***magnitude*** of mass and in particular, since our particles are viewed as distortions of the *fabric of spacetime*, how the nature of that fabric in turn influences particle mass. To begin with, the answer to that question appears straightforward: Eq. (25-7), which we repeat here for reference,

$$m = \left(\frac{4A}{3c^2}\right)\left(N\sqrt{WR}\right), \tag{26-1}$$

indicates that it is simply the magnitude of ***energy linear density***, multiplicative parameter A, that is of concern. From a practical point of view, however it is more revealing to use its inverse, $G = c^4/4\pi A$, for which we have a time-honored value (although measured at a *scale* many orders of magnitude ***larger*** than that of interest here) in which case we have

$$m = \left(\frac{c^2}{3\pi G}\right)\left(N\sqrt{WR}\right) \tag{26-2}$$

or, setting $W = N\Delta r$ with $\Delta r = L_P$, i.e. equal to the Planck length, often cited as the smallest value for which the laws of physics can reasonably apply,

$$m = \left(\frac{c^2}{3\pi G}\right)\left(N^{3/2}\sqrt{RL_P}\right). \tag{26-3}$$

At this point we embark upon an interesting numbers game: suppose we consider the proton's parameters and the established value for the Planck length. That is we use

$$m = m_P = 1.673 \times 10^{-27}\,\text{kg}$$
$$R = R_P \simeq 2 \times 10^{-15}\,\text{m},$$
$$\mu = 1,$$

for the proton and the value of $L_P = 1.62 \times 10^{-35}\,\text{m}$ for the Planck length. Furthermore, we make the numerator of Eq. (23-9), also as *small* as possible by setting $N = 1$. Then, if we also insert the established value of $G = G_0 = 6.673 \times 10^{-11}\,\text{m}^3/\text{kg} \cdot \text{s}^2$ into Eq. (26-3), we find the value of m_P as ***computed*** per that equation to be disconcerting: it is more than eight orders of magnitude *too* ***large***! Which means that G is eight orders of magnitude too ***small*** and, correspondingly, A is also eight orders too ***large***.

On the other hand, the fact of the matter is that we have no empirical information as to what G might be at the ***scales of interest***. Thus, it is meaningful to compute what it ***ought*** to be, given the particle parameters listed above for the proton (except for particle width which we parametrize using dimensionless N rather than W). That is, we compute

$$G = \left\{ \left(\frac{c^2}{3\pi m} \right) \sqrt{\mu R L_{Pl}} \right\} N^{3/2} \tag{26-4}$$

with the result shown in the Log–Log plot of Figure 26-1. We see a straight line where N ranges from about 10^2 to 10^8 and G is shown as the dimensionless ratio G/G_0 ranging from about 10^{30} to 10^{40}.

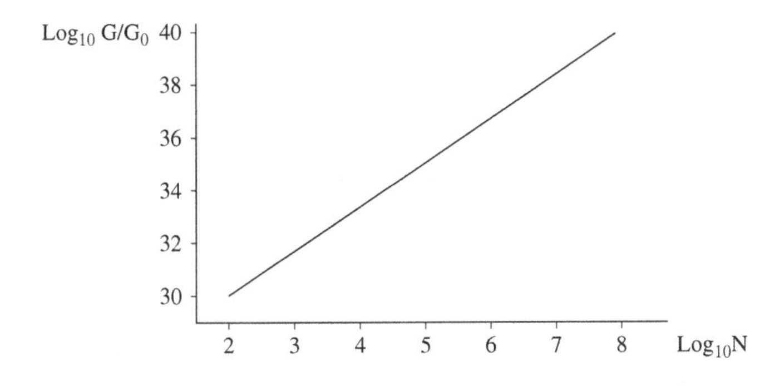

Figure 26-1.

A very interesting point is that for $G/G_0 = 10^{39}$ which puts G in the same range of values as the SM's so-called ***Strong Force***. It occurs for $N \simeq 1.6 \times 10^7$, at which point particle width is $W = NL_P \simeq 2.6 \times 10^{-28}$ m, which is still a million times smaller than the "size", R_e, of the electron as computed in the previous section (rather skinny MS). There is thus the suggestion, here, that whereas $G = G_0$ is usually associated with the ***gravitational attraction*** between ponderous bodies in the ***macroscopic*** world, given a greatly *increased* **local** value it can also be held responsible for the ***strong force*** which mediates ***close-in*** interparticle interactions. Correspondingly, the associated *local* value of A, the ***linear energy density*** that our particles encounter according to Eq. (23-9) is greatly ***reduced***.

In other words, our solitonic particles operate in what amounts to a *local toroidal* ***trough***, a situation strikingly reminiscent of the symmetry-breaking topology that preconditions the ***Higgs mechanism*** invoked by the Standard Model to impart mass to its particles out with an important difference we'll talk about later. A final note before we proceed to enlarge upon that observation: We should add that, per the relation between size and mass found in the previous section, the plot applies irrespective of particle size. Furthermore, the dependence on particle width (per Eqs. (25-7) and (26-1)) is not unique to our model; at some point in its development string theory was realized to encompass extensions to higher dimensional versions including *membranes* for which coupling strength also increases with width [48].

27

A Parallel to the Higgs

In this chapter we follow up on the above suggestion that each basic AM particle experiences "a situation reminiscent of the symmetry-breaking topology that preconditions the ***Higgs*** mechanism". For purposes of comparison we need something definitive but the complete Higgs mechanism is, necessarily, rather complicated; without it the Standard Model would not be able to attribute mass to the elementary particles of the model without violating some of its basic theoretical underpinnings, primarily SU(2) gauge symmetry and the efficacy of renormalization. However, it is not necessary to invoke the entire mechanism here because, in the first place, SU(2) symmetry in the AM is ***inherent*** to particle structure and, furthermore, renormalization is not an issue. What is ***basic*** to the SM's Higgs mechanism, however, is the need to introduce a symmetry-breaking version of the ***potential energy*** that each particle encounters. That potential is also of interest here because of the way in which symmetry breaking was introduced into the AM, that is, by way of the initial *implicit* ***assumption*** of toroidal topology and the ***reduction***, as per the above, of the value of linear energy density, A, of spacetime in the neighborhood of the MS.

For ***comparison*** purposes, at this point, we adopt a version of the Standard Model's Higgs potential that corresponds to the commonly-invoked **Landau–Ginzburg** model, the one based on the spontaneously broken symmetry encountered in ferromagnetism [49], namely

$$V = \mu^2|\phi|^2 + \lambda|\phi|^4, \qquad (27\text{-}1)$$

where $|\phi|$ is the amplitude of an unspecified, complex field variable, and seek an extremum of the total potential function V as per the usual procedure. We obtain

$$\frac{dV}{d\phi} = 2\mu^2|\phi| + 4\lambda|\phi|^3 = 2\phi(\mu^2 + 2\lambda\phi^2) = 0 \tag{27-2}$$

for the first derivative and

$$\frac{d^2V}{d\phi^2} = 2\mu^2 + 12\lambda\phi^2 \tag{27-3}$$

for the second derivative. As usually discussed, when μ^2 is positive we see that the potential is concave upward with a minimum of $V = 0$ at $|\phi| = 0$. However, when μ^2 is negative, there is a maximum of $V = 0$ at $|\phi| = 0$ and a minimum at

$$|\phi| = \sqrt{\frac{-\mu^2}{2\lambda}} \tag{27-4}$$

which puts the potential function into the desired spontaneously broken symmetry form basically as shown in the well-known shape of Figure 27-1 where the vertical axis is the potential function, V, and the horizontal axis is $|\phi|$. Qualitatively, the initial descent is dominated by the quadratic term in Eq. (27-1) and the upward return by the quartic term. Clearly, the two terms in the potential function must be of different algebraic sign with the quadratic term negative.

So much for the Standard Model's side of the comparison. The Alternative Model side is a radically ***different way*** to arrive at a Higgs-like potential. But before we display it, we can gain some

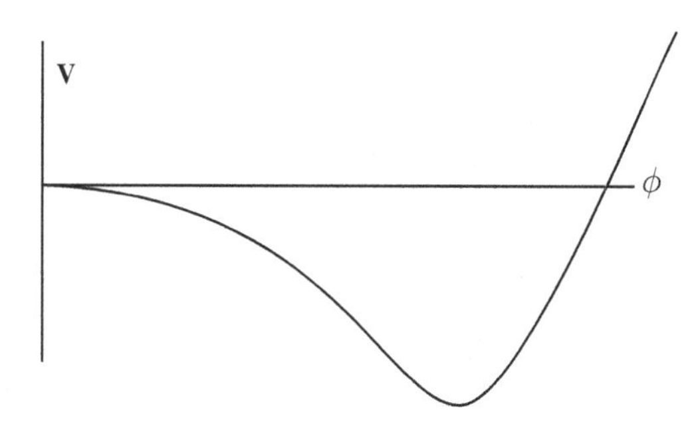

Figure 27-1. Higgs potential as per the Standard Model.

additional insight if we consider the following very short but *not at all inconsequential* ***Theorem*** (which we shall, therefore, refer to as the NI Theorem):

Suppose we have two variables, x and y which interrelate according to each of two equally-valid relationships;

$$\begin{aligned} &1. \quad y = ax^n, \\ &2. \quad y = bx^{-n}. \end{aligned} \tag{27-5}$$

Summing the two, multiplying by x^n and "shuffling" we obtain the equally valid relationship,

$$V \equiv -b = -2yx^n + ax^{2n}. \tag{27-6}$$

Seeking extrema, we set

$$\frac{dV}{dx} = -2n(yx^{n-1} + anx^{2n-1}) = 0 \tag{27-7}$$

which is satisfied by $x = 0$ in which case, $V = 0$, and also by

$$x = \left(\frac{y}{a}\right)^{1/n}, \tag{27-8}$$

in which case, to decide between maxima or minima we compute

$$\frac{\left(\frac{d^2V}{dx^2}\right)}{2n} = x^{n-2}[(2n-1)ax^n - (n-1)y] - x^{n-1}\frac{dy}{dx}. \tag{27-9}$$

Using Eq. (27-8), i.e. $x = (y/a)^{1/n}$, we find for the second extremum that

$$\frac{\left(\frac{d^2V}{dx^2}\right)}{2n} = (n-1)ax^{2n-2} \tag{27-10}$$

which is ≥ 0 for $n \geq 1$ (actually for all n because the two relationships just change places for negative n) which signifies that the relationship $x = (y/a)^{1/n}$ obtains at a ***minimum***, at which point, $V = -ax^{2n}$. But, we note that this is **relationship #1**, the one associated with the trough we discussed in the previous chapter. And, furthermore, the maximum at $x = 0$ jibes with **relationship #2**; All are very reasonable.

So now, going back to Eq. (27-6), setting $n = 2$ and making the associations

$$\begin{aligned} b &\leftrightarrow -V, \\ x &\leftrightarrow |\phi|, \\ 2y &\leftrightarrow \mu^2, \\ a &\leftrightarrow \lambda, \end{aligned} \tag{27-11}$$

we see that the formalism we have obtained from our little NI ***theorem*** and the formalism of the ***Higgs potential***, usually described as patterned after magnetization theory, are ***equivalent***. We might, in fact, go so far as to characterize the Higgs potential as just a special case — an example — of our little NI Theorem! (Why not? Let's do it!)

Of course, we have an ulterior motive here — what we are really aiming at is a way to use the theorem as a method of generating a ***similar formalism*** with our particle model and we now explore an approach toward that goal. To begin with we note that, with $n = 2$, the ***first*** relationship in the theorem obtains if we set $y = R$, $x = m$ and, as derived from Eq. (26-3) in Chap. 6,

$$a = \frac{9\pi^2 G^2}{c^4 N^3 L_P}. \tag{27-12}$$

For the ***second*** relationship we have adopted an expanded approach to the mass versus size issue, one involving considerations of ***quantum*** mechanical behavior, albeit of the "old quantum theory", in the expectation of a fuller quantum mechanical treatment at a later date. We refer to what in Sec. II was called "Einstein's ***Ansatz***", the one that maps the quantized trajectory of a particle moving in a central field of force onto a torus in order to resolve momentum ambiguities. Actually, the plan here is to apply the Ansatz in ***reverse***; that is we ***postulate*** the existence of a central field of force that acts upon the solitonic distortion we described above, the implication for our case being that there exists a putative centrally located ***source*** for such a field and, correspondingly, discrete orbits that might be identifiable with the toroidal loci of our model's electron and nuclei.

In other words, we shall now consider the simultaneous ***validity***, as in the above, of ***two***, completely independent models for our electron and nuclei, one the ***solitonic*** model discussed in the foregoing and the other an ***orbital*** quantum model, hereinafter to be known as the "Converse Einstein/Bohr (***CE/B)*** model. As we know, 100 years ago, Niels Bohr formulated a model of the Hydrogen atom that predicts that element's discrete line spectra. Bohr considered the electromagnetic attraction

$$F_E = \frac{e^2}{R_o^2} \tag{27-13}$$

between a proton in the nucleus and an electron in orbit as balanced by the centrifugal force

$$F_C = m_e \frac{v^2}{R_o} \tag{27-14}$$

experienced by the electron due to its orbital velocity. Bohr also postulated that angular momentum is quantized in multiples of Planck's constant according to

$$mvr = n_o \frac{h}{2\pi}. \tag{27-15}$$

The combination of these relationships results in an expression for the ***radius*** *of the orbit*, namely

$$R_o = \frac{h^2 n_o^2}{4\pi^2 m e^2}. \tag{27-16}$$

For the case of our (CE/B) model we should expect a ***similar*** relationship, except that the force of attraction would not be electromagnetic and, of course, we are concerned with much smaller scales. As we have seen, solitonic distortion in our particle model is concentrated in a small, mobile region of longitude to which we now attribute the *totality* of particle mass. Although we have not yet discussed its kinematics, assuming that the Bohr analysis is applicable implies that the implicit ***velocity*** of circulation of such a "pseudo particle" is ***immaterial***. However, if the force of attraction between the putative central "entity" (we are not yet prepared to call it a particle) and our solitonic pseudo particle were ***gravitational*** rather than electromagnetic, we should replace the force, F_E, by the ***gravitational*** force

$$F_G = \frac{GMm}{R^2}, \tag{27-17}$$

where M is the mass of a putative centrally located entity with the result that the relationship for "***orbital***" radius R becomes

$$R = \frac{h^2 n^2}{4\pi^2 GMm^2} \tag{27-18}$$

and amounts to just replacing me^2 by GMm^2.

We now assume, as in the above, the simultaneous validity of our two models, the solitonic behavior as per Eq. (26-3), repeated here for reference,

$$m = \left(\frac{c^2}{3\pi G}\right) N^{3/2} \sqrt{RL_P}$$

and that for the CE/B model, namely Eq. (27-18), or, more compactly,

$$\begin{aligned} &1. \quad R = \alpha m^2, \\ &2. \quad R = \frac{\beta}{m^2}, \end{aligned} \tag{27-19}$$

where with $n = 1$

$$\alpha = \frac{9\pi^2 G^2}{c^4 N^3 L_P},$$

$$\beta = \frac{h^2}{4\pi^2 GM} = \frac{h^2 A}{\pi c^4 M},$$

which is also ***in the form*** of our little NI theorem! In other words, we now know how to realize what we might now refer to as the Alternative Model's ***version*** of a ***Higgs*** potential. There are various ways to express it. One way, in analogy to Eq. (27-1) that we wrote down for comparison purposes, is as the potential

$$V_{comp} = -\eta^2|\phi|^2 + \varsigma|\phi|^4, \tag{27-20}$$

where, to make things a little more *physical* we have used

$$\begin{aligned} A &= \frac{mc^2}{2g}, \\ \eta^2 &= \frac{4Rg^2}{c^4}, \\ \varsigma &= 4\alpha\left(\frac{g^2}{c^4}\right)^2, \\ g &= \sqrt{Rr}. \end{aligned}$$

To recapitulate, the main thesis of our Alternative Model of the elementary particles has been that it constitutes an actual, *physical* ***manifestation*** of the taxonomy, interactions and attributes of the Standard Model but without recourse to the latter's quarks, gluons or color. At some point in its development, the SM was considered by its participants to be a complete theory with one notable exception; it was impossible to endow its particles with mass and retain gauge invariance and renormalization efficacy unless the symmetry of the potential energy they experience is broken. Ergo, the notion of a potential energy field — the so-called Higgs field — such that the potential energy a particle experiences depends on its location as measured in the ***coordinates of that field***.

We emphasize that such coordinates are ***not physical*** — that is they do not have the dimensions of ***space***. As noted at the beginning of this section, the formal expression for the field is sometimes described as being modeled after the symmetry-breaking field of ferromagnetic magnetization theory. The field may be characterized by its "hill and valley" ***topography*** — that is, a centrally located mound surrounded by a radially-symmetrical trough. As we have seen, it consists of a quadratic term that describes the descent down from the hill and a quartic term that reverses the downward slope and climbs up out of the valley. Symmetry breaking occurs because location at the top of the hill is only conditionally stable and the particle's preferred location in field coordinates is ***somewhere*** at or near the bottom of the valley but in ***any radial*** *direction.*

The identical field is assumed to *exist at* ***any location*** *throughout space* so that a particle moving through it will be affected in a manner sometimes rhetorically described as a kind of drag on its motion. It seems fair, at this point, to say that the Higgs field of the SM may be viewed primarily as an ***attribute of space*** rather than of any individual particle, even though not all particles are affected by it to the same extent. Finally, the culmination of a rigorous search, using the particle accelerator at CERN to collide protons together at high energy has apparently identified a hitherto unknown particle which has the attributes associated with the quantum of the Higgs field, the so-called Higgs boson, thus, it is maintained, removing the last obstacle to the complete validation of the Standard Model.

What has emerged in the foregoing, and in particular in this section, is that a *real, physical* ***parallel*** to the hill-and-valley topography of the Higgs field can be associated with each elementary AM ***particle*** by permitting the gravitational field in its ***interior*** to vary in a particular way in ***particle coordinates***. The Higgs-like characteristic was arrived at herein by assuming the simultaneous validity of two independent relationships between particle mass and size: one associated with our sine–Gordon ***solitonic*** MS and the other Niels Bohr's model of the *Hydrogen atom* as adapted to a ***gravitational*** rather than an electromagnetic force. Recall, if you will, our repeated insistence that what may be described as the "last" for our sine–Gordon, ***Möbius strip*** elementary particles, namely an actual ***torus***, *does* ***not*** *really exist.* What we see instead is the friendly ***valley*** of this gravitational characteristic providing the necessary ***toroidal*** *topology* for our MS to ***occupy*** while the ***hill*** provides a putative *central* ***field*** *of force*, not initially considered as part of the model, for the MS to take shape as per Einstein's Ansatz.

28

The Central "Entity"

Admittedly, what we have described above is a rather unusual approach especially since, we note, not only was the underlying toroidal topology of our AM particles initially simply *assumed* without explanation or justification, now we are introducing a further hypothetical entity. However, now that we have gone this far, suppose for the sake of argument that we go one step further: suppose that we simply ***assume the existence*** of a complete Higgs-like characteristic interior to each elementary particle *to begin with.* This is essentially analogous to the SM's approach to the Higgs field, eventually justified, it is now claimed, by the discovery of the long-sought Higgs boson. In other words, the novel hill and valley topography of the AM is then to be viewed as a fundamental attribute of each elementary particle.

That being the case, our purpose in this section is to learn something about our hypothetical "entity" with mass M, the one responsible for our putative central field of force. On the basis of Eq. (27-19), and the associated expressions for α and β, its mass can be written in two ways, either as

$$M(R) = \frac{c^4 h^2 (N^3 \mu L_{\mathrm{P}})}{36\pi^4 G^3 m^4} \tag{28-1}$$

or as

$$M(R) = \frac{\left(\frac{9}{4}\right) h^2 G}{c^4 (N^3 \mu L_{\mathrm{P}}) R^2}. \tag{28-2}$$

Interjection. Note that both expressions above contain c, h and G, the speed of light, Planck's constant and the gravitational "constant", respectively, quantities related to Relativity, Quantum Mechanics and Gravitation, where the latter two quantities were introduced in connection with the Bohr formula as modified in terms of gravitation rather than electromagnetic attraction. We shall have occasion to discuss the implications thereof in Sec. VII.

Logically, as long as we postulate the existence of at least a formal parallel to the SM's Higgs field we really ought to acknowledge the existence of a particulate AM analogy to its quantum, the Higgs boson, especially in the light of that entity's apparent discovery at the CERN particle accelerator. But given that our Higgs field exists in real space interior to each elementary particle there ought to be a real ***manifestation*** of such an entity in that region and what better candidate than the ***central hill*** of our hill and valley topography. The question is then what that boson has to say about that feature's attributes, and, in particular its mass. To investigate this point we begin with a formula based on combining Eqs. (28-1) and (28-2) for the mass of the putative hill (that is we multiply and take the square root!):

$$M = \frac{\left(\frac{h}{2\pi}\right)^2}{GR_n m_n^2}, \tag{28-3}$$

where G is the gravitational constant and R_n and m_n are nucleon size and mass.

The ratio of hill mass to that of the Higgs boson is then

$$\frac{M}{m_H} \equiv \gamma_H = \frac{\left(\frac{h}{2\pi}\right)^2}{GR_n m_n^3 \gamma_{H/N}}, \tag{28-4}$$

where $\gamma_{H/N}$ is the ratio of the mass of the Higgs boson to that of a nucleon which is apparently about 125. The first thing we notice, in analogy to what we discovered in the "numbers game" of Sec. III is that setting G equal to the nominal value G_0 makes γ_H too big while setting it equal to, say, $10^{39}G_0$, the value associated with the valley, makes it too small. So, again, apparently what we ought to do is find out what it *ought* to be in order to make $\gamma_H = 1$. As normalized to G_0 we then have

$$\frac{G}{G_0} = \frac{\left(\frac{h}{2\pi}\right)^2}{G_0 R_n m_n^3 \gamma_{H/N}} \tag{28-5}$$

whereupon, using

$$\begin{aligned}
\frac{h}{2\pi} &= 1.06 \times 10^{-34}\,\mathrm{J \cdot s}, \\
G_0 &= 6.67 \times 10^{-11}\,\mathrm{m^3/kg \cdot s^2}, \\
\gamma_{H/n} &= 1.25 \times 10^2, \\
R_n &\simeq 2 \times 10^{-15}\,\mathrm{m}, \\
m_n &\simeq 1.67 \times 10^{-27},
\end{aligned}$$

we find $G/G_0 = 1.44 \times 10^{35}$ to be the desired value. That makes it 6944 times ***smaller*** than the value associated with the *valley* (10^{39}) which was calculated, it will be recalled, to make nucleon parameters match their accepted values. Correspondingly, it also means that our hill, as expressed in terms of linear energy density parameter A, is 6944 times ***higher*** than our valley, a comfortable situation.

The message to this point is that our Alternate Model can now claim not only to manifest the taxonomy, interactions and attributes of the Standard Model but that we now have an expanded ontological model of the elementary particles that includes what amounts to a physical ***manifestation*** of a Higgs field and, at least formally, a candidate for the Higgs boson. However, again we emphasize, where the SM's Higgs field is an attribute of the space through which particles move, the AM's "Higgs field" is an integral part of and, thus, a manifest attribute of ***each elementary particle*** even as the particle moves ***solitonically*** through space.

From a larger point of view, the AM's "Higgs field" may be considered to constitite a hitherto-neglected fundamental entity required for particle formation, an integral, ***necessary*** feature of our particle model. Note that, initially, our Möbius strip (viewed as a concatenation of torus knots) was ***assumed*** to exist by virtue of its topology for purposes of analysis, said analysis then leading to the solitonic formalism that validated the assumption. Now, in this enhanced point of view, our *solitonic* MS exists ***in concert*** with a Higgs-like internal gravitational field, one which, in fact, provides what amounts to the toroidal region ***within which*** the solitonic nature of the MS plays out. In other words, our basic particle model now consists of what might be called a symbiotic combination of those two entities.

A final comment: in the same spirit that our formulation for the Alternative Model's epistemology constituted an amalgamation of the contributions of Sakata and Gell-Mann/Zweig, we note here that our "indigenous parallel" to the Higgs potential involves what might be characterized as a "***rapproachement***" between two old friends and sometime philosophical adversaries, Albert Einstein and Niels Bohr, at least insofar as their contributions to said parallel are concerned.

29

A Constraint

Going back to Eqs. (28-1) and (28-2) of the previous chapter, we note that since both of these relationships refer to either the electron or a nucleon, we are really concerned with four cases, namely

$$
\begin{aligned}
M(m_n) &= \frac{\left(\frac{k_m}{m_n^4}\right)}{G^3} \\
M(m_e) &= \frac{\left(\frac{k_m}{m_e^4}\right)}{G^3} \\
M(R_n) &= \left(\frac{k_R}{R_n^2}\right) G, \\
M(R_e) &= \left(\frac{k_R}{R_e^2}\right) G,
\end{aligned}
\qquad (29\text{-}1)
$$

where

$$k_m = \frac{c^4 h^2 (N^3 \mu L_\mathrm{P})}{36\pi^4}$$

and

$$k_R = \frac{9h^2}{4c^4(N^3 \mu L_\mathrm{P})} = \frac{\left(\frac{h}{2\pi}\right)^4}{k_m}.$$

Figure 29-1 shows a representative quartet of these relationships separated into two pairs of inclined parallel lines. Those that descend to the right are labeled $M(m_n)$ and $M(m_e)$ and those that rise to the right are labeled $M(R_n)$ and $M(R_e)$. Due to the large span of values to be considered, these lines are displayed in log–log form with the independent variable being $\gamma = \operatorname{Log} G/G_0$.

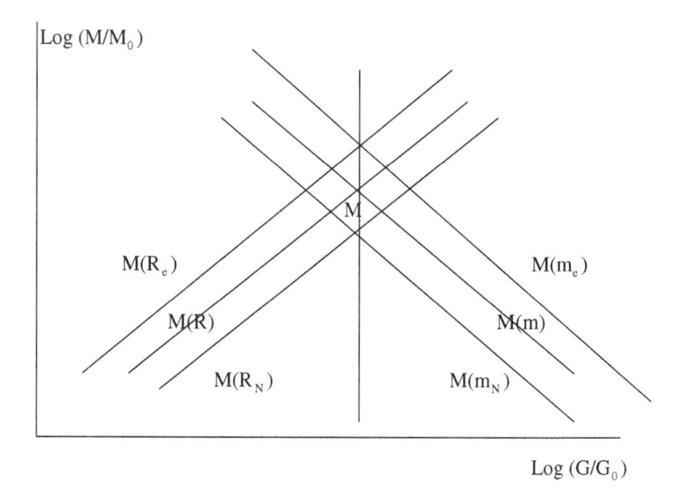

Figure 29-1. Normalized mass vs. normalized gravity.

As a consequence of the basic solitonic relationship between mass and size, both pairs of lines are separated by the same logarithmic increment in the vertical direction — that is, the direction in which Log (M/M_0) is measured with M_0 being an arbitrary reference. We also see two additional lines, one located midway between each pair — i.e. it bisects the strip defined by each pair. Since each such bisector constitutes the arithmetic mean of the bisected pair in the Log–Log plot, it represents the geometrical mean of their actual values. In other words, we have that

$$
\begin{aligned}
M(\beta_m) &= \sqrt{M(m_n)M(m_e)} \quad \text{and} \\
M(\beta_R) &= \sqrt{M(R_n)M(R_e)},
\end{aligned}
\tag{29-2}
$$

where the βs represent the bisectors.

Finally, we also see that the two pairs of lines intersect so as to form a quadrangle centered about the intersection of the two bisectors. Again, since the two pairs of lines are separated by the same vertical increment the upper and lower intersections of the lines lie on a vertical line whose midpoint is the intersection of the bisectors.

The mass of that point, as expressed in logarithmic notation becomes

$$4\text{Log}\, M = \text{Log}\, M(m_n) + \text{Log}\, M(m_e) + \text{Log}\, M(R_n)\text{Log}\, M(R_e) \tag{29-3}$$

or, as a geometrical mean,

$$\begin{aligned}\overline{M} &= \frac{\sqrt{k_m k_R}}{(Gm_n m_e \sqrt{R_n R_e})} \\ &= \left(\frac{1}{Gm_n m_e \sqrt{R_n R_e}}\right)\left(\frac{h}{2\pi}\right)^2 .\end{aligned} \tag{29-4}$$

Using the solitonic relationship between mass and size, Eq. (29-4) translates into

$$\overline{M} = k\left(\frac{\frac{h}{2\pi}}{m_n m_e}\right)^2, \tag{29-5}$$

where $k = (c^4/9\pi^2)(N/G)^3(L_{\text{P}})$, which tells us what the (geometrical mean of the) masses of the putative central entities would amount to given only the masses of the two kinds of elementary particle. Conversely, if we assume the reality of that entity and attribute a value to it, i.e. the number M, then rewriting Eq. (29-5) gives us

$$m_n m_e = \left(\frac{k}{\overline{M}}\right)^{1/2}\left(\frac{h}{2\pi}\right) \tag{29-6}$$

which is the equation of a set of hyperbolas parametrized by N, G and M.

However, under further assumption that all the quantities on the R.H.S. are fixed, what we have is a ***constraint*** between the mass of the electron and that of a nucleon. And, of course, we know at least one set of quantities, namely those that give the measured values for m_n and m_e. In other words, given the mass of either particle, the mass of the other is automatically specified. Of course, this expression in and of itself says nothing about a tolerance to associate with departures from the constraint, nor of the ***ratio*** of one mass to the other. Generally speaking, also, the constraint has to do with particles as they exist rather than with the processes that produce or modify them (although if either is known to have emerged first in the history of the universe, the mass of the other is also specified).

There is, however, a notable exception: in the Alternative Model (and only therein), neutron decay, as discussed in the foregoing, constitutes a (literally) graphic illustration of the constraint in action as the original neutron goes through several configurational changes including the ***expansion*** of one particle (the "protonucleon") and the corresponding ***contraction*** of another (the "protoelectron") to their final sizes as proton and electron, respectively. These modifications therefore occur as required by the equivalent size version of the ***constraint***, namely

$$R_n R_e = \left(\frac{k'}{\overline{\overline{M}}}\right)\left(\frac{h}{2\pi}\right), \tag{29-7}$$

which follows using the previously employed relationship between mass and size and with actual implementation resulting from the inclination of the AM's version of the proton to favor torsion over curvature and that of the electron being the opposite. We'll talk about this matter further in Sec. VII where we explore some general asociations.

30

Postscript: Spinoza, Clifford, Einstein, Wheeler and Me

Benedict (Latinized from Baruch) Spinoza (1632–1677), a contemporary of Leibnitz, was one of the most important philosophers of the 17th century (but, of course, you already knew that). What is of interest here is what may be termed his ontological philosophy of existence, known as Monism, namely that, while it can evidence itself in various ways, there is ***only one***, universal, absolutely infinite, fundamental ***substance*** (which he equates with God, a viewpoint whose complexity we need not address here). To which I might legitimately claim a kinship for my (repeated insistence that there is only) ***one field*** ontology, ***spacetime*** itself. You may recall our discussion of reductionism as an epistemological tenet; here, we have a rather significant reduction (from, say, the field-per-particle ontology of the SM)!

About Clifford, we have already displayed in some detail his solitonic notion of matter as a distortion of the geometry of space. Einstein, you may recall, first appeared quite early in this section where we highlighted his "***ansatz***", the idea that allowed momentum ambiguities associated with motion in a central field of force to be resolved by mapping onto the surface of a torus. Subsequently, of course, we reversed the ansatz idea, ***postulating*** a gravitational central field of force *a la* Bohr to produce what we called an indigenous parallel to the Higgs potential.

All of which is just a matter of recalling what we have talked about in the preceding. However, in terms of new things to talk about, Professor John Archibald Wheeler (July 9, 1911–April 13, 2008) is our man and, in that regard, the first thing we want to do is to apologize for the somewhat cavalier way in which we treated his pithy (there is that word again!) summary of General Relativity because that was not exactly the way he felt about the situation at all. Professor Wheeler was a gentleman and a scholar, a brilliant physicist, known for his originality and imagination. Once he turned to a study of GR, Wheeler delved into the subject in great depth, as was his wont. In terms of the underlying philosophical implications therein he was influenced by both Spinoza and Clifford. Also, since he taught at Princeton for many years, he was a good friend of Einstein's and had many conversations with him. In his autography [50] he recounts one such that took place not long before Einstein's death and he says that "the vision that animated Einstein throughout his later years — a vision I had many occasions to discuss with him — was a vision of a totally geometric world, a world in which everything was composed ultimately only of spacetime".

That last phrase "composed only of spacetime" sounds, on the face of it, a lot like the solitonic particles that are featured in this book! As is often the case, however, the way things sound "on the face of it" is misleading. What Wheeler eventually came up with to fit that phrase was called a ***Geon***, a circulatory wave phenomenon held together in orbit by the gravitational influence of its own energy density. Wheeler's first design involved electromagnetic waves in circular orbit but eventually he also considered gravitational waves in both circular orbits and spherical shells. He was quite intrigued by the whole concept for which he came up with another pithy (sorry) saying, "mass without mass". Of course there were problems and, to make a long story short, the main one was that Geons are only provisionally stable. To quote Wheeler himself "— the slightest disturbance will cause a Geon to collapse or dissolve, radiating away its energy to the cosmos."

Actually, again on the face of it, the instability is somewhat predictable because, for one thing, Geons are ***not solitons*** (like our Torus knots or MS). In fact when Wheeler asked Einstein, himself a legendary innovator, what he thought about them, Prince Albert replied that he thought they might be unstable. I found out about Geons at a meeting of physicists — I do not remember which but it is one of those where you get ten minutes to describe an hour's worth of radical concepts to people who, for some strange reason, are mainly interested in their own ideas, radical or not. That was the meeting where I was told by a professor of physics that my Alternative Model particles are "Just Geons" which is patently false. That was also about the time I decided to forego attendance at further such meetings; they were a waste of time.

However, I ended up buying Wheeler's book (see above) and learning a lot, including the meaning of ***Geometrodynamics***. My conclusion is, let's forget the Geon saga: the idea of Geometrodynamics was another successful and important Wheeler innovation in a long productive career of innovating, teaching, mentoring and just generally being one of the most productive physicists of the 20th century. In their admirable 1995 book on Gravitation and Inertia [51], Ciufolini and Wheeler devote the first 172 pages to "Einstein Geometrodynamics" and the word otherwise pops up throughout the book! The idea was to encapsulate for relativity theory the essentials "of spacetime geometry — dynamic, changing geometry, influenced by mass, capable of propagating, and in turn, influencing mass." In other words, "as a companion to the old, familiar ***Electrodynamics***, a word used to describe the theory of electricity and magnetism — how electric and magnetic fields are created, propagate, and get absorbed."

In any event, I would like to think that this section (Sec. V) has contributed some not insignificant additions to the general subject of Geometrodynamics. Right up front of the list of potential candidates for significance is just the nature of particles as solitons with toroidal topology. The connection between particle mass and size is another as is the resulting estimate for the size of the electron. Then there are a set of items that stem from the notion that gravitation might vary in the interior of a particle (since AM particles have interiors!), the first one being a potential connection to the so-called "strong force" between nucleons which, in the Standard Model, is sort of a residual of the basic interaction posited to exist between quarks.

And finally, there is the combination of our originally-postulated solitonic behavior with an additional postulation (as per my little "not inconsequential" theorem) of a central source of gravitational activity as per a modification of Bohr's model of the Hydrogen atom, coupled with Einstein's mapping of particle orbits onto a torus. This had several consequences, the first, of course, being what I called an "indigenous parallel" to the SM's Higgs field. But if you recall that "Interjection" I interjected a little while ago (we were talking about the nature of the putative "central entity") the use of the Bohr theory inserted an additional element into the proceedings: at that point our particles became explicitly associated with quantum behavior because Planck's constant was involved in the formalism. One result was the description in the preceding of a "hyperbolic constraint" relating the masses (or equivalently the sizes) of the electron and the nucleons. There may be more such but I do not have any ready-made at this time so this is a good time to close this section. What is coming up is exciting — at least it got me excited writing it — and, admittedly, more than a little speculative here and there.

VI
Assorted Topics, Some Speculative

"*So long as a man imagines that he cannot do this or that, so long as he is determined not to do it; and consequently so long as it is impossible to him that he should do it*"

(*Benedict (Baruch) Spinoza,*
Theological Political Treatise, 1670)

"*He who hesitates is lost*"

(Author Unknown)

"*Look before you leap*"

(Unknown — but probably another — author)

"*Leaping is for the young.*
But don't give up on imagination"

(Jack Avrin)

31

Time: CPT, Instantons and Triplication

The moving finger writes, and, having writ,
Moves on: nor all your Piety nor Wit
Shall lure it back to cancel half a line,
Nor all your tears
Wash out a word of it.

(the Rubaiyat of Omar Khayyam; translated by Edward Fitzgerald)

Many papers, essays, tracts, even books have been written about the nature of time and from many points of view, although none quite so lyrical as those few lines by Omar via Fitz. Our concern in this chapter is much more limited: you may recall a statement made at the end of Sec. II to the effect that our FMS are to be regarded as occupying a $2+1$ spacetime wherein the ***out of plane*** dimension is ***time***. Thus, the discrete charges of the quirks are associated with ***steps*** in time. In this section we explore some of the ramifications of that point of view. To begin with, we note that the underlying ontology of our alternative model may be viewed as a kind of **Kaluza–Klein** theory, especially in terms of Klein's modification of Kaluza's five-dimensional modification of General Relativity that combines it with electromagnetism. Apparently what Klein did amounts to ***compactifying*** the fifth dimension of Kaluza's original theory into a circle in spacetime [28]. Correspondingly, we note what amounts to the ***meridianal*** compactification of our solitonic knots so as to encircle their toroidal "last", also in spacetime with time as the dimension normal to the plane of the torus and ultimately of the flattened Möbius strip, the FMS.

More on CPT

We shall return to these considerations a bit later in this chapter but first, we consider some *explicit* ways that time enters into our deliberations. For example, we talked about CPT (charge, parity and time) invariance in Sec. IV but here's another way to demonstrate the detailed invariance of our particles: consider a quirk in isolation and the accompanying reference coordinate system. Parity here implies the orientation of the direction of traverse as portrayed in the down quirk of Figure 31-1 by the arrows, counterclockwise, as shown is (+) and clockwise is (−). We will consider only rotations by 180° about each of the portrayed axes. Axis x which represents *time*, here, is to be interpreted as pointing out of the plane of the diagram.

1. It is reasonably clear by inspection that rotations around the x (time)-axis — that is in the plane of the diagram — have no effect on either C, P or T.
2. A rotation around the y-axis produces Figure 31-2.

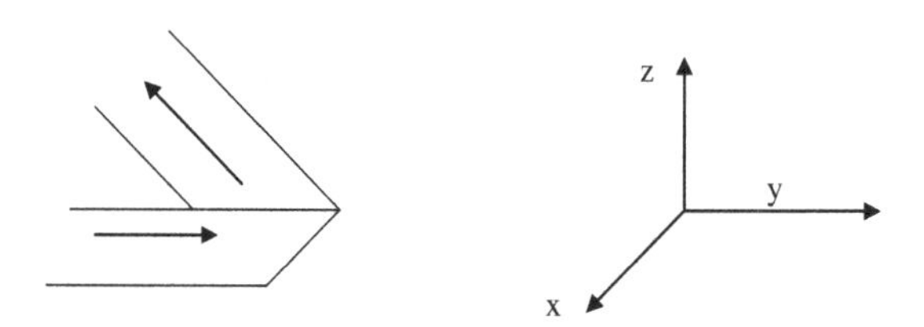

Figure 31-1. Down quirk reference.

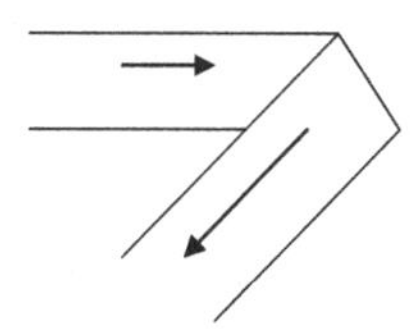

Figure 31-2. Rotation about the y-axis.

We see that the attributes C and P have changed but not T; time remains as defined. Thus

$$
\begin{aligned}
&\mathrm{C} \rightarrow \mathrm{C}^* = -\mathrm{C}\\
&\mathrm{P} \rightarrow \mathrm{P}^* = -\mathrm{P}\\
&\mathrm{T} \rightarrow \mathrm{T}
\end{aligned}
$$

so if we associate the value of −1 to a *changed* attribute (as per the rotation) and +1 to an unchanged attribute, we see that the products $\mathrm{C^*P^*T}$ and CPT are equivalent, i.e. the rotation produces no change in the CPT product.

3. A rotation about the z-axis produces Figure 31-3 and, upon inspection, we find exactly the same result: the new product is again $\mathrm{C^*P^*T} = \mathrm{CPT}$.

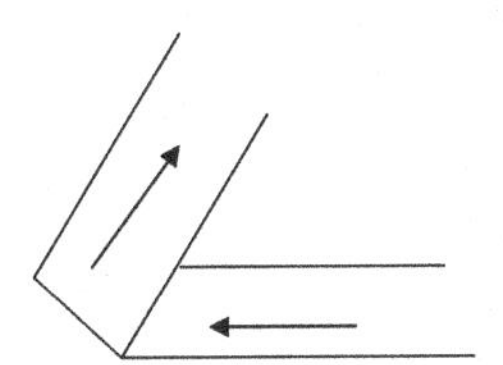

Figure 31-3. Rotation about z-axis.

We conclude that the CPT product of a quirk, and therefore of an FMS, is ***invariant*** to orientation in ***spacetime***. And, a *companion* conclusion to these deliberations, the Wheeler–Feynman notion that an ***antiparticle*** looks like the associated ***particle*** moving *backwards* in time is demonstrable by considering, for example, the converse of Figure 31-1, a d* quirk, which we portray simply by reversing the traverse arrows as shown in Figure 31-4.

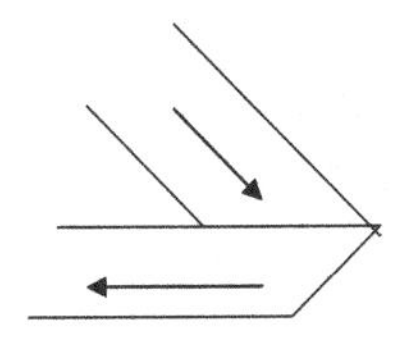

Figure 31-4. Antidown quirk.

Then, considering what that figure would look like as seen from the *back* of the diagram, namely as shown in Figure 31-5; we see that it is also a d quirk which validates the (really important) W-F characterization.

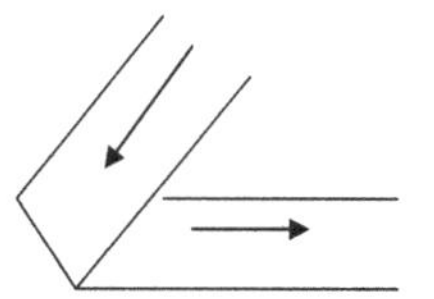

Figure 31-5. Same as seen from "back".

Instantons

Another ramification of our treatment of time is the relationship of quirks to ***instantons***. The treatment here is strictly heuristic; we consider ***four*** kinds of instanton characteristics, labeled I, AI, RI and RAI in a two-dimensional (1+1) spacetime as shown in Figure 31-6. Although the mathematical considerations are more involved, in essence, the prototypical instanton characteristic labeled I here, is just a ***step*** change between two eigenstates (which we label here as residing in the ***spatial dimension***) that occurs in a very short interval of ***time***, and the label AI stands for the anti-instanton characteristic with a step in the opposite direction [52].

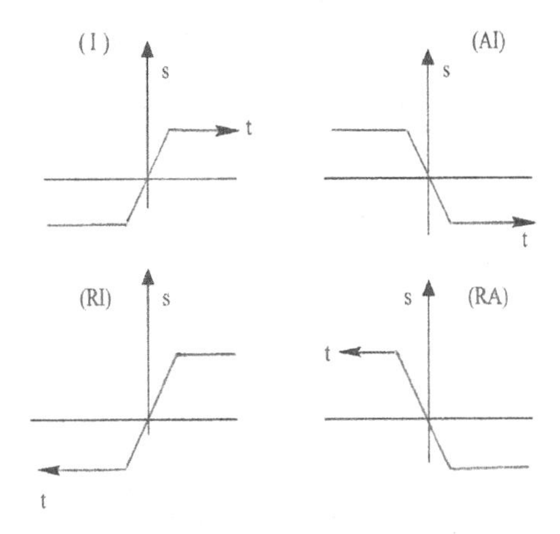

Figure 31-6. Four instantons characteristic (as per this book!).

Customarily, only the ***two*** characteristics, I and AI, are discussed; you knew that, right? Well, characteristics RI and RAI are introduced, here, strictly for reasons of *symmetry*, RI to portray an instanton progressing in negative time and, similarly, RAI to portray an anti-instanton progressing in negative time. In any event, Figure 31-6 shows all four of them; I kind of like the way that looks; and besides, that's what this section is supposed to be about according to its heading!

Here we are showing time as flowing horizontally and space vertically. Now if you think about it, a quirk is just the ***converse***, a transition region, a jump from one region of time to another in a very small region of ***space*** as we show in Figure 31-7 with the same four diagrams *relabeled* such that the time and space axes are reversed but space is measured along the direction of ***traverse***.

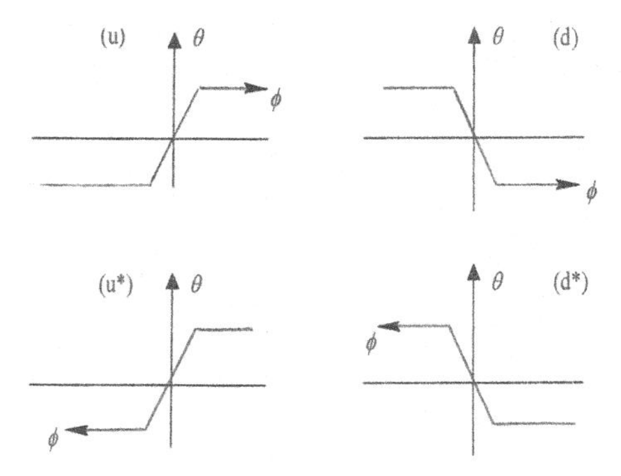

Figure 31-7. "Correspondences" of quirks and instantons.

These four diagrams thus portray the two basic ***quirks*** and the corresponding ***antiquirks***. That is we have the correspondences

$$\mathrm{I} \rightarrow \mathrm{u},$$

$$\mathrm{AI} \rightarrow \mathrm{d},$$

$$\mathrm{RI} \rightarrow \mathrm{u}^*,$$

$$\mathrm{RAI} \rightarrow \mathrm{d}^*.$$

In other words, our basic set of quirks and antiquirks is isomorphic to the above set of instantons and reverse instantons.

One implication to be drawn from the foregoing is that, to the extent that our model of the basic fermions reflects reality, the postulated instanton characteristics, RI and RAI, would do so as well (and conversely). Another way to view the relationship between Instantons and FMS is depicted in Figure 31-8 where we first show the four members of Figure 31-6 but with time running vertically in the picture to match the way we have shown the quirks. Then, in Figure 31-9 we show each quirk as a combination of two of those four

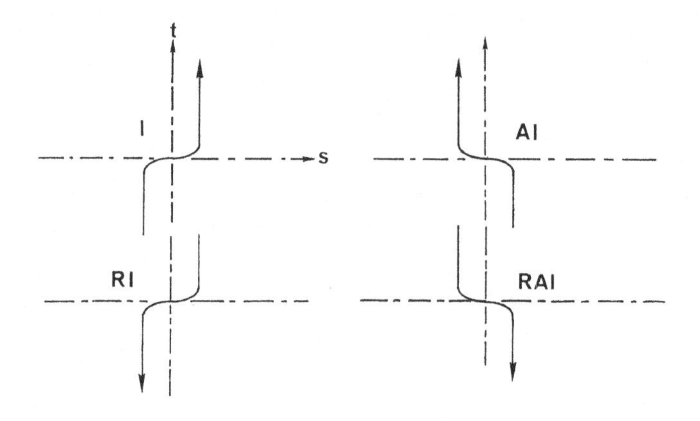

Figure 31-8. Instanton characteristics rotated 90°.

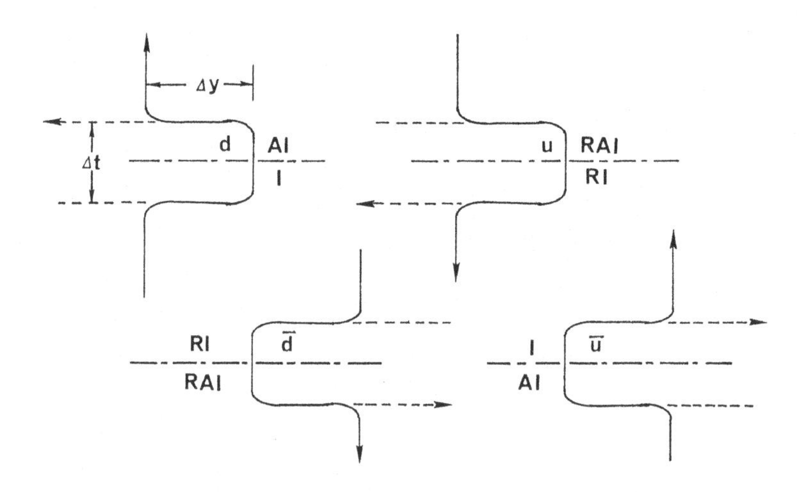

Figure 31-9. Building quirks from instantons.

but taking place in small intervals of both space and time, with space measured again along the traversal direction. Now, if we let both those temporal and spatial increments decrease in the right way, in the limit we have a spatial ***delta*** function of time for each quirk. Or, better, we can view space along the traverse direction as the spatial derivative of its own instanton characteristic. Thus we have

$$\mathrm{d}(s) = \left(\frac{ds}{dt}\right) \mathrm{I}(t),$$

$$\mathrm{d}^*(s) = \left(\frac{ds}{dt}\right) \mathrm{RI}(t),$$

$$\mathrm{u}(s) = \left(\frac{ds}{dt}\right) \mathrm{RAI}(t),$$

$$\mathrm{u}^*(s) = \left(\frac{ds}{dt}\right) \mathrm{AI}(t).$$

Triplication of Time

Finally, I introduce, here, another ***speculative*** notion, namely that there actually is ***more*** *than one dimension* of ***time*** as well as the three-space. The allusion at the beginning of this chapter connecting our modeling to the ontology of **Kaluza–Klein** theory, suggests that what we are considering here amounts to an extension of the K-K notion. We recall that we left our icoshedral geometry in Sec. VI, with three mutually orthogonal duadic planes and an associated set of ***orthogonal axes*** that define the geometry of the icosahedron. We know that these define the icosahedron as a ***framework*** for the three generations of the family. Beyond that, however, suppose we view each of the three ***duadic*** planes as a kind of ***internal space***, each peculiar to interactions that take place ***within*** and ***only*** within a corresponding generation, one plane to a generation. For example, in duad UD as per Figure 20-6, the first generation might be *exemplified* by the interactions discussed therein. This is in contrast to the quirk-replacement, *intergenerational* particle groupings shown on the *triangular* planar scaffolds in that section and the sets of particles that occupy the loci emplaced thereon; clearly *their* role is to summarize and organize particle *family* structure.

However, there is still the associated ***orthogonal*** *coordinate* system to consider and, in that regard, we recall once more, our identification of the out-of-plane coordinate for individual FMS with ***time*** as in the immediately preceding sections. Here we have ***three*** duadic planes, as per the above, "one per generation". Since, in the sense posited above, a given *duadic plane* uniquely constitutes the internal space for the FMS of a *corresponding* generation, the next step follows by implication: we simply associate a unique "***time***" ***label*** to the particular coordinate vector associated with (that is, ***normal*** to) *each* ***duadic plane***. That is, we would associate ***unique unit time*** vectors to the normals to planes UD, CS and TB, respectively. Thus, the spacetime of reference — that ***inhabited*** by the particles of our model — that spacetime becomes a $\mathbf{3+3}$ (rather than a $3 + 1$) dimensional continuum (however, see below).

Of course, spacetimes of more than four dimensions are nowadays commonplace but, being space-like, the extra dimensions must be "hidden from view" in a "Calabi–Yau" manifold in some way, shape or form as in string theory. Our "triplication of time is more like an ***upgrading*** of the familiar time coordinate of our macroscopic existence into a time ***manifold*** with the triplicated time coordinates forming its orthogonal basis. In effect, *time* is ***promoted*** to be on the same *dimensional* footing as *space*. However, since ***real*** time would not necessarily correspond to any of the three orthogonal coordinates, we might construct something like a linear combination of coordinate vectors. Or, expanding upon that notion, we would construct the equivalent of a **Kobayashi–Maskawa** (KM) [53] matrix, one that produces transition probabilities between generations, in which case we would need to establish a correspondence to the latter.

One way to approach such matters is in terms of the algebra of ***quaternions***, the rationale being that from the point of view of, say, a "fiber" traversing one of our MS/FMS, there is only *one dimension of space* that matters; in effect, this "*traversal*" spacetime is $(1+3)$-dimensional, an entity that lends itself to that kind of treatment. Quaternions were introduced in Sec. III where we found that a quaternion can be expressed as

$$\mathbf{P} = p_0\lambda_0 + \mathbf{p}, \tag{31-1}$$

where

$$\mathbf{p} = p_1\lambda_1 + p_2\lambda_2 + p_3\lambda_3,$$

and

$$\begin{aligned} \lambda_0\lambda_0 &= 1, & \lambda_1\lambda_2 &= -\lambda_2\lambda_1 = \lambda_3, \\ \lambda_0\lambda_i &= \lambda_i\lambda_0 = \lambda_i, & \lambda_2\lambda_3 &= -\lambda_3\lambda_2 = \lambda_1, \\ \lambda_i\lambda_i &= -1, & \lambda_3\lambda_1 &= -\lambda_1\lambda_3 = \lambda_2. \end{aligned}$$

We then found the outer product of two such entities to be expressible as

$$\mathbf{P} \otimes \mathbf{Q} = (p_0 q_0 - \mathbf{p}\cdot\mathbf{q}) + (\mathbf{p}_0 q + q_0\mathbf{p}) + \mathbf{p}\times\mathbf{q}, \tag{31-2}$$

where

$$\mathbf{p}\cdot\mathbf{q} = p_1q_1 + p_2q_2 + p_3q_3$$

and

$$\mathbf{p}\times\mathbf{q} = (p_2q_3 - p_3q_2)\lambda_1 + (p_3q_1 - p_1q_3)\lambda_2 + (p_1q_2 - p_2q_1)\lambda_3$$

or as a 4×4 matrix, the direct product of a column and a row vector (as in Sec. III to form boson matrix M) as

$$(p_0, p_1\lambda_1, p_2\lambda_2, p_3\lambda_3)^{\mathrm{T}} \otimes (q_0, q_1\lambda_1, q_2\lambda_2, q_3\lambda_3)$$

$$= \left\| \begin{matrix} p_0q_0 & p_0q_1 & p_0q_2 & p_0q_3 \\ p_1q_0 & -p_1q_1 & p_1q_2 & p_1q_3 \\ p_2q_0 & p_2q_1 & -p_2q_2 & p_2q_3 \\ p_3q_0 & p_3q_1 & p_3q_2 & -p_3q_3 \end{matrix} \right\|. \tag{31-3}$$

Also, we found that we can subdivide this matrix so as to summarize and highlight its intrinsic organization as a 2-dimensional manifestation of the content of Eq. (31-2). Thus, Figure 31-10, shows a *scalar*, an *inner product*, *two vectors* and a *cross product*, the latter split between two pieces of matrix real estate.

If we delete all the terms containing p_0 or q_0 in Eq. (31-3) we are left with a 3×3 matrix, the direct product of the 3-vectors

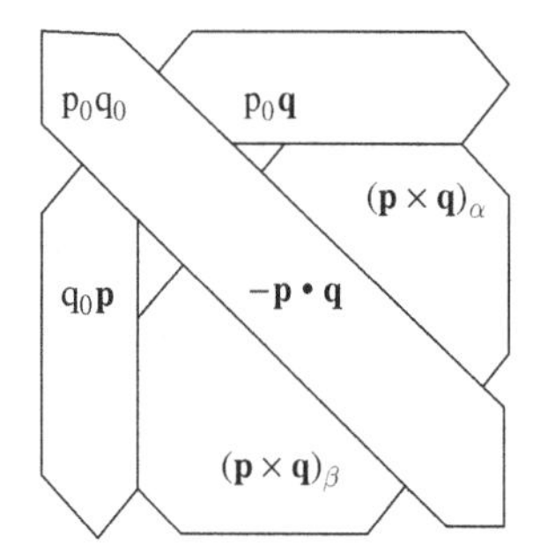

Figure 31-10. Breakout of quaternion product (repeat).

p and **q** or alternatively, the direct sum of the inner product and cross product of these vectors as shown in Figure 31-11. From the quaternion point of view, vectors **p** and **q** are also known as pure quaternions and the 3×3 matrix is a 2-dimensional manifestation of their outer product. In any event, whether in full or reduced form, we have an ***operational*** representation of the outer product of two quaternions, which can be used to operate on additional quaternions.

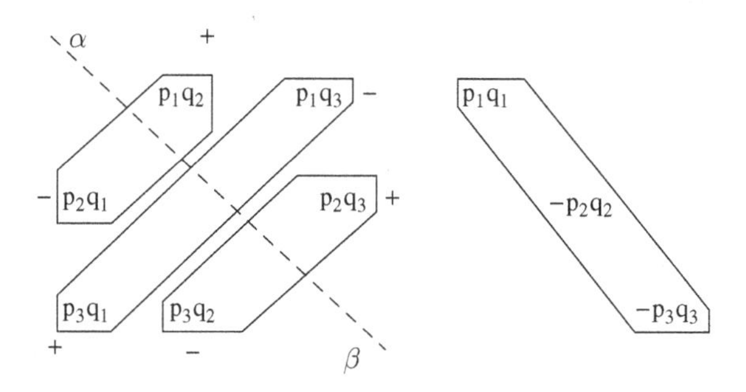

Figure 31-11. Breakout of inner and outer products (repeat).

From the point of view of our $(1+3)$-dimensional "traversal" spacetime, the analog to Figure 31-10 is Figure 31-12 which reflects the definition of a 4-dimensional vector $\mathbf{R} = s + \mathbf{T}$, where $\mathbf{T} = t_1\hat{\imath} + t_2\hat{\jmath} + t_3\hat{k}$ (with the t's being the equivalent of the λ's) is a 3-dimensional "time" manifold and s is the single space dimension — just the reverse of our usual $(3+1)$-dimensional spacetime.

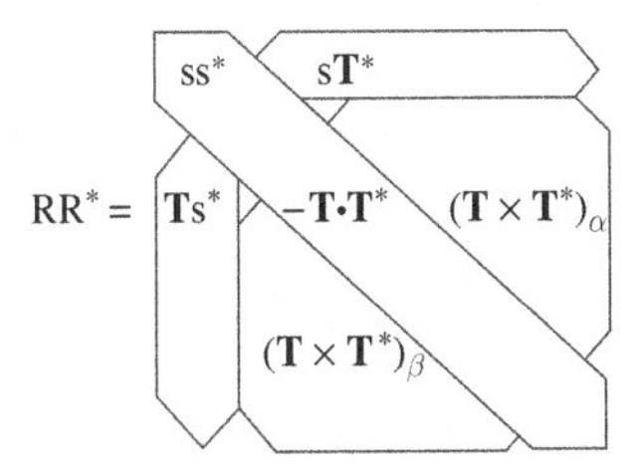

Figure 31-12. Breakout of four-dimensional quaternion product.

The scalar and vector terms involve space only and spacetime interaction, respectively; deleting them leaves us with the inner and outer products that occupy the following 3×3 matrix, the direct product of the *triplicated time vector* and its conjugate:

$$\mathbf{T} \otimes \mathbf{T}^* = \begin{Vmatrix} t_1 t_1^* & t_1 t_2^* & t_1 t_3^* \\ t_2 t_1^* & t_2 t_2^* & t_2 t_3^* \\ t_3 t_1^* & t_3 t_2^* & t_3 t_3^* \end{Vmatrix} . \tag{31-4}$$

At the same time, we recall that the Standard Model employs the Kobayashi–Maskawa (KM) matrix [53] to systemize intergenerational interactions.

$$V = \begin{Vmatrix} V_{ud} & V_{us} & V_{ub} \\ V_{cd} & V_{cs} & V_{cb} \\ V_{td} & V_{ts} & V_{tb} \end{Vmatrix} . \tag{31-5}$$

Thus, if we identify indices $i = 1, 2$ and 3 of the t_i with u, c and t, (up, churm and top) respectively and $j = 1, 2, 3$ of the t_j^* with d, s and b, respectively, we see that the two matrices are isomorphic.

Also, as noted in [53] a useful approximation to the KM matrix is

$$V = \begin{Vmatrix} 1 & \lambda & \lambda^3 \\ -\lambda & 1 & \lambda^2 \\ \lambda^3 & -\lambda^2 & 1 \end{Vmatrix} , \tag{31-6}$$

where $\lambda = s_{12}$ and $s_{ij} = \sin\theta_{ij}$, the angle being the "mixing angle" between the ith and the jth generations. Thus our $t_1 t_2^*$ and $t_2 t_1^*$ terms correspond to the KM terms λ and $-\lambda$, respectively, of the first and second generations, as they should, and our $t_2 t_3^*$ and $t_3 t_2^*$

terms correspond to the KM terms λ^2 and $-\lambda^2$ of the second and third generations, as they should.

We conclude that there is a demonstrable ***correspondence*** between the time triplication approach and the KM matrix approach to systematizing the triplication of generations on the family tree. This is about as far as this notion has progressed and there are of course problems that quickly come to mind. One such is that, empirically, the three analogous SM generations are very unequally experienced; for example, it takes a lot more energy to find (what is ***interpreted*** in the SM to be) a top quark than it does a down quark. Thus, the three-dimensional symmetry that our time triplication manifold inherits from the icosahedral structure that gave rise to it in the first place must be *broken*. How this is to be accomplished has not yet been addressed in any detail although a speculative application of time triplication with broken symmetry is discussed in Chap. 32, below. Also, validating the notion of time triplication would introduce a radical new symmetry into fundamental physics, a *generalization* of the symmetry of spacetime over and above the point of view that usually underpins relativity (either special or general).

32

Dark Matter

What is known nowadays as the Standard Cosmological Model postulates the existence of "Dark Matter", the source of the gravitational influence that keeps galaxies from flying apart. Most literate people have at least seen or heard of it and are aware that it is quite unlike "ordinary matter" with which it otherwise does not interact. And, perhaps also, that there are numerous candidates for its constituents. We will talk some more about DM when we get to cosmology in Sec. VII, but for now, here is ***another*** candidate, admittedly speculative, but emerging from our alternative model in a logical way: Always lurking at the perimeters of our earlier deliberations but never really contributing to them is that enigmatic fourth elementary fermion, the one with a twist of +3 and a charge of +2. Clearly, this FMS as well as its taxonomical combinations have little to do with the alternative model's taxonomy or interactions. Nevertheless, they must appear in our modeling by virtue of symmetry and so they do, at each level of fusion, a kind of ***super (anti) symmetry***.

That being the case, the question naturally arises as to whether our super(anti) symmetric particles might actually have as much relation to reality as do elementary fermions A, B, C and their taxonomical combinations. Now you may be aware of the current (as of this writing) search for evidence of what is known as supersymmetry (no "anti"), that is, the expansion of Standard Model elementary particle taxonomy to include the existence of a bosonic partner for each fermion and, vice versa, a fermion for each boson.[1] In fact, its apparent absence to date seems to pose a threat to some of the SM's basic tenets, including a relationship to particle mass and, more to the point of this chapter, the source of Dark Matter [54].

Well, let us assume that **our** dual particles actually do ***exist*** but that, in the manner of Dark Matter, we cannot "see" them simply *because* they do not interact, at least in any kind of electromagnetic way, with normal matter. Though they are therefore "dark", they should be endowed with mass, possibly very heavy mass, as inferred from the apparent gravitational influence that led to their putative existence in the first place, in which case the ***nature*** of that putative endowment is the issue. One possibility is simply that, as per the relationship of particle mass to size generated in Sec. V, particle D is just larger than the others in which case we would like to know why and wherefore, which gets us into the general question as to why some particles with mass and not others exist, something we will touch upon in Sec. VII.

Or, we could invoke the notion, discussed in Chap. 31 of the ***triplication*** of time as part of the explanation for additional generations of particles. Although, conceptually, such triplication should exist at every point of space, there appears to be no prior reason for a "*dark*" particle's ***ordinary*** time to be the same as it is for ***our*** ordinary taxonomical world. That is, it could instead be one of the other orthogonal time coordinates in the triplication, perhaps the one associated with our ***heaviest*** generation in which case finding a "dark" particle might be very difficult.

Of course, all of this is pure speculation but here's another idea that will make the above seem the height of conservatism! To begin with, perhaps you have also seen the recent paper in Scientific American [55] about the way in which the matter we know and love, that is to say "ordinary" matter — let us call it "***our*** matter" — is distributed in space. The paper discusses how simulations of the

[1] And they are all to live together happily ever after. (I may be getting a little punchy here.)

universe, using all "known" aspects of how it evolves, end up with dark matter spreading across space in a network of filaments — "a cosmic web" — at the intersections of which galaxies, especially dwarf galaxies, of "***our*** matter", at least those we see, ***tend*** to form (there are complications which is what the paper is about, but that are of no immediate interest to us).

Actually the notion of the cosmic web of filaments is not new but in any event, here is my idea: I am going to ***assume*** that the universe we encounter is, in reality a ***composite*** of ***two***, ***complementary***, "Component Universes": one involving "***our*** matter" and one "***their*** matter", with "their" translating as "***dark***" for us. That is it; two component universes; *Ours* and *theirs*; *Light* and *dark*; *Yin* and *Yang*! As per the thesis of this book, ***Ours*** is built from the elementary fermions A, B and C and their conjugates as well as the taxonomy that results from their combination. What I postulate here is that ***theirs*** is similarly built from the fermions D, C and B and the associated combinations. The two components influence each other gravitationally — obviously dark matter influences where the light matter congregates as per the above.

However, although, as we observe it right now, the two universal components coexist, I further ***postulate*** that they may not always have done so or will do so; they have different ***histories*** — Cosmic "life cycles" as they say. In fact, one or the other may, in some sense, be in an ascendant phase and the other decaying. For example, let us suppose that the dark component has been around much ***longer*** than our light one. The implication of course is that the ***former*** was the host for the emergence of the ***latter***, a situation that would appear to answer the question as to what came ***before*** the Big Bang! Furthermore, it would render ***moot*** the either-or relationship between the steady state and big bang scenarios for our universe; the answer is ***both***! But of course it is the dark component that presently dominates and determines the underlying ***structure*** of *our* universe and, if our hypothesis is correct, did so for a long time, even before our component existed. And, not unlike electricity and magnetism, the two components taken together constitute one, complete entity. And as far as we are concerned, the dark component is indispensible for the continued existence of our universe, at least in the form that permits our existence!

At any rate, "apparently", (meaning the Standard Cosmological Model's best current estimate) about 14 billion years ago, what we have been calling herein ***our*** universal component began. Most of us are reasonably familiar with the SCM's Cosmic Life Cycle story; a monstrous "inflationary" phase, followed by "ordinary" expansion with the emergence of particles, background radiation, stars, galaxies, etc., — there's little to be gained by recounting it at this point. However, there is one more point that ought to be stressed: the implication of what we have been saying here is that not only did our component begin at a particular ***epoch***, it must have begun at a particular ***location*** in the ***already existing dark*** component universe. Of course there is no way we can specify what that location is; all we can say, in other words, is that our component universe began at a ***particular spacetime*** point in an ***already existing*** spacetime. That may be a fine distinction but it differs from the usual point of view that, in essence, time and space grew, along with the expansion of the universe as we know it; after all, we are not the sole arbiter of that matter. And a word of caution: location of the point of emergence in the ***Dark*** universal component does ***no***t mean that we can discern a similar location in ***our*** component universe; being part of it forbids that!

And, just so you don't get the wrong idea, in some sense, what I have been describing is not, of course, really new in its entirety; the notion of continual emergence of new universes from pre-existing old ones has been around for some time, having originated with a number of cosmologists including **Vilenkin** and **Andrei Linde** [56], also one of the pioneers of Inflation theory. Nevertheless, it is hereby asserted that this chapter has introduced a new wrinkle into the grand and glorious fabric of cosmological speculation. It would of course be nice to have a convincing sequence of mathematical argumentation to lend credence to the wrinkle but, again, maybe next time.

A couple of final remarks: first, both of the above universal components are dynamic entities and, although at this point we don't understand it, the interplay between them includes, one would expect, some kind of oscillatory process that we'll talk about a bit in Sec. VII. And last but not least; the observation that the two universes are complementary entities arising, inevitably, in our initial formalism is discussed as part of a broader point of view in Appendix F; please don't miss it.

33

TQFT

Connecting the alternative model to "dark matter" would be most gratifying but there are more "shovel ready" topics for additional development. One such is the proposition that our model constitutes a manifestation of Topological Quantum Field Theory (**TQFT**). According to Wikipedia, "A topological quantum field theory is a quantum field theory which computes topological invariants". If that is a little too cryptic, it goes on to say that "in a topological field theory the correlation coefficients do not depend on the metric on spacetime That means that the theory is not sensitive to changes in the shape of spacetime; if the spacetime warps or contracts, the correlation functions do not change. Consequently they are topological invariants."

That sounds pretty clear and it would appear that from that top-level point of view, considering our Alternative Model as a TQFT is eminently justified: in terms of the differential geometrical point of view as per Sec. V, an MS is a soliton, a ***topologically-quantized*** disturbance in and of the fabric of spacetime. However, there is more than one approach to the subject and, in fact, more than one way to characterize what it means and since **Witten's** work in developing the notion of a TQFT [57] and his groundbreaking interpretation of the **Jones** polynomial for knots in that light [58] are well known as is **Atiyah's** axiomatic approach to the formalization of the subject [59] we shall consider both as definitive. A comprehensive discussion of the Alternative Model's role as a TQFT must await further publication, for which we apologize but the work of both authors is too technical for this book and we are not sure we can do justice in translation as we write the book!

However, we can summarize an ***approach***, one we adopted in preliminary unpublished work as follows: With reference to Figure 33-1, the idea behind the *upper* branch is to assert that MS/FMS ontology, especially as employed in the development of the taxonomy that results, embodies the fundamental Atiyah requirements for a TQFT as detailed in Atiyah's monograph. In other words, what has emerged from a detailed comparison is that on that basis, the model does indeed qualify as a legitimate TQFT.

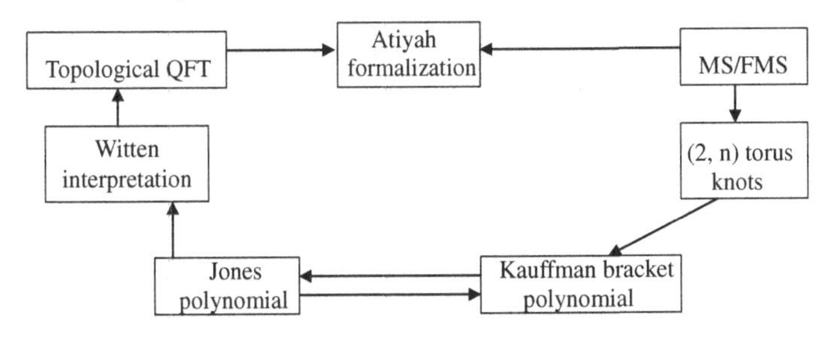

Figure 33-1. Top level alternative model/TQFT relationships.

But there are other encouraging indications to that effect. In particular, as stated in the Atiyah frontispiece; "The material presented here rests primarily on the pioneering work of Vaughan Jones and Edward Witten, relating polynomial invariants of knots to topological quantum field theory in 2 + 1 *dimensions*." Which, fortuitously, is also the dimensionality of the FMS in our alternative model — an encouraging start, at least qualitatively. Also pertinent is the Atiyah statement: "The Jones polynomial has been generalized in a variety of ways." And "— [a] fundamental way involves choosing a compact Lie group G and an irreducible representation." — "The original Jones polynomial corresponds to taking G = SU(2)–." In that regard, we note that, in making a necessary choice of pertinent gauge group, Witten was also led to select the group SU(2). Thus, our alternative model appears to coincide with the ***parametric basis*** of **Witten/Jones/Atiyah** theory at least in what has been described as its simplest form.

With regard to the lower branch of Figure 33-1, the equivalence between the ***Jones*** polynomial and the **Kauffman** *bracket* polynomial (colloquially, the "Bracket") is well known [26] as, in particular, is the Bracket of (2, n) torus knots that we discussed in Sec. III. So is the relationship between such knots and the MS as discussed in this book (An MS may be viewed as a concatenation of (2, n) torus knots or as a framed, two-strand braid with closure). Thus, given the putative validity of Witten's interpretation of the Jones polynomial as a TQFT, the path is closed and the implication is that no further discussion is necessary; the FMS model is, ipso facto, a TQFT.

Although that logic leaves out the demonstration of a *common* ***ontology***, a primary reason for the emphasis on the upper branch, it turns out that the existence of a correspondence between ***composite*** FMS and what are known in TQFT as composite ***cobordisms*** (derived from the notion of "common borders") emerges in the detailed analysis of Atiyah's axioms as mentioned above. This is a very ***important*** concept, in my view ***the*** most important such, in a TQFT and, in fact, Atiyah's axiomatic approach appears to emphasize it. Here is my abstracted list of the axioms with a considerably modified language and including only those items deemed pertinent to this chapter. It should be understood that this is strictly our interpretation put together without consultation with the author.

1. A topological QFT in dimension d assigns a complex vector space $Z(\Sigma)$ to each compact, oriented, smooth, d-dimensional manifold Σ and a vector $Z(Y) \in Z(\Sigma)$ for each compact, oriented $(d+1)$-dimensional manifold Y with ***boundary*** Σ ([59, p. 12]).
2. As per the above, in the Jones–Witten theory, $d = 2$ so that Σ ***is a surface***.
3. $Z(\Sigma^*) = Z(\Sigma)^*$, where Σ^* denotes Σ with the opposite orientation and $Z(\Sigma)^*$ is the dual space [59, p. 12].
4. There is also the notion of a "***composite cobordism***", two $d+1$ manifolds, say Y_1 and Y_2, united at a ***common boundary*** by the pairing between the dual spaces $Z(\Sigma)$ and $Z(\Sigma^*)$ [59, p. 13].
5. For a closed $(d+1)$ manifold, the vector $Z(Y)$ becomes a complex number, an invariant of the manifold which "— can be computed

from any decomposition $Y = Y_1 \cup_\Sigma Y_2$" [59, p. 13]. When $d = 2$, $Z(Y^*) = \overline{Z(Y)}$, the complex conjugate.

Items 4 and 5 of the axioms talk about the ***composition*** and ***decomposition*** of $d + 1$ manifold ***cobordisms*** in terms of their ***union***. When $d = 2$, Atiyah's common border is by definition a ***vector*** space. The analogous activity in our alternative model is the ***fusion*** and ***fission*** of FMS for which the sum of the number of half twists (NHT) of the component FMS equals the NHT of their combination. Figure 33-2 then shows the explicit analogous situation, for the fusion of a d quirk with its conjugate d*. We see the border as, in this case, a rectangular $(1 + 1)$-dimensional space (δt in this case is the very small step in time that occurs at a quirk) with vectors that coalesce when the borders actually merge. (Imagine one of the two figures rotated so that the rectangular borders coalesce and the large arrows are continuous.)

Two other items, not actually listed as axioms in the reference should also be listed as pertinent here:

6. In order to actually produce invariants for links/knots in 3-manifolds, it is necessary to populate the $(d + 1)$ manifold with oriented links. Slicing a composite cobordisim slices link L and creates a boundary Σ at which $\partial L \subset \Sigma$ is a collection of signed points. Signing amounts to marking each point with either $a+$ or $a-$ sign and the vector space $Z(\Sigma)$ now becomes

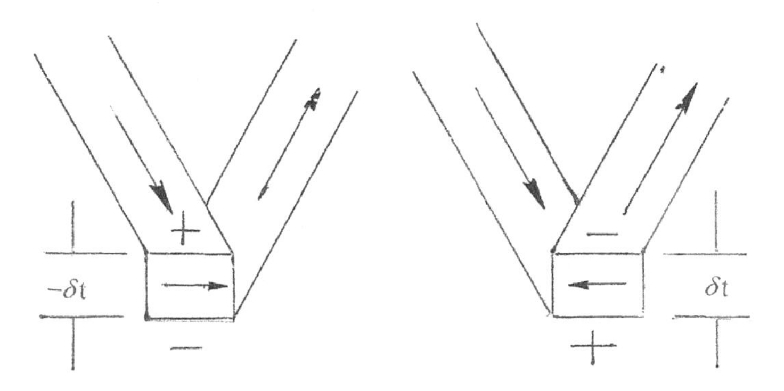

Figure 33-2. AM analogs to composition and decomposition cobordisms.

explicitly associated with the markings. Again this is depicted in Figure 33-2, for our model as interpreted from the reference [59, p. 15].

7. Finally, for "mathematically technical reasons" (e.g. to "pin down scalar factors" and eliminate "phase ambiguities") the Witten formalism is found to be incapable of actually producing Jones polynomials unless the links are "***framed***" so that they are no longer one-dimensional. Again we note this inherent feature of the relationship between the MS in our Alternative Model and the associated knots or braids.

Recapitulation

In other words: as per the discussion in the Introduction, the lower branch of Figure 33-1 summarizes the logical connection between the visualizable model under discussion and the Witten demonstration of how topological quantum field theory and knot theory are related. However, since the lower branch lacks an ontological demonstration of that relationship, the upper branch, involving the Atiyah formalization of the Witten Jones connection, was emphasized in the preceding. The strategy was to encapsulate the items of the formalization relevant to the AM and then proceed to demonstrate the relationship. That demonstration is summarized below:

Item 1. This item is partially satisfied by the existence of a binary vector space, the $1+1$ surface formed by a quirk as illustrated in Figure 33-2. The associated toroidal $2+1$ FMS manifolds complete the correspondence.

Item 2. Note that $d = 2$ as per item 1.

Item 3. The surfaces of Figure 33-2 for basic fermions correspond to the Σ of this item. Reversal of the arrows as per the basic antifermions corresponds to the Σ^*.

Item 4. Fusion of a fermion and an antifermion forms the "common boundary" uniting Σ and Σ^*.

Item 5. The invariance of twist and charge and their calculation applies here.

Item 6. Our torii are populated by links/knots as per the nature of MS/FMS discussed above. Signed points become evident at the boundaries upon the fission of a higher order FMS that reveals the kind of surface shown in Figure 33-2.

Item 7. As discussed early on in the book, framing of links/knots is automatic for MS/FMS.

Short of a detailed analysis, the foregoing should give the flavor of the relationship of our Alternative Model and Atiyah's axiomatization of Witten's approach to uniting TQFT and knot theory.

34

The String Theory Connection

This chapter is going to be even sketchier than the previous three; String Theory is a really big subject and there is no way we can do justice to it in summary here. However, we can mention a few items that tend to indicate some commonality with our alternative model. For example, we should emphasize that each of the concatenating torus knots that makes up an MS is explicitly wound around a torus which amounts to what is known in ***string theory*** as ***toroidal compactification***. But, of course, there is more to the comparison than that. For one thing, the coefficient of the kinematic term in the Lagrangian yielding the sine–Gordon equation as discussed in Sec. V is identifiable as ***string tension***. Also, string theory is notable in developing a number of versions, each requiring extra spatial ***dimensions*** which must then be hidden from view in the **Kaluza–Klein** [29] tradition. The hidden dimensions then take on what are known as **Calabi–Yau** shapes [60] (after the mathematicians who invented them) some exotic in the extreme as the number of dimensions increases. However, the simplest such are for two dimensions in which case the Calabi–Yau shape is simply a ***torus***!

Also, as is well-documented, string theory has developed into something considerably more encompassing, often referred to as ***M theory*** [61] with a historic shift in attention from 10 to 11 dimensions and a number of theoretical "manifestations". However, there appears to be something like a "core" M theory, associated specifically with gravitons, objects known as five-branes, and, what is of more interest herein, ***two-branes***, i.e. membranes, the strength of whose coupling depends on their width. Which, of course, implicates the MS of our alternative model as we saw explicitly in Sec. V.

On the other hand, there are some fundamental differences. One is scale: where our nucleons subtend about 10^{-15} m of space and, as calculated in Sec. V, our electrons about 10^{-22} m, string theory objects exist down around the Planck scale of 10^{-35} m. Of course, this does not really matter to the ***basis*** of string theory which is that "real life" objects are modeled as different frequencies of vibration of the strings, a radically novel departure from most of the physical world heretofore!

While string theory has scored some notable accomplishments, for example in terms of connections to the nature of Black Holes and an inherent inclusion of what purports be the model of a graviton, it continues to be the target of criticism from its major competitor, Loop Quantum Gravity (see Chap. 41) for the title of the most promising road to a theory of "Quantum Gravity" namely that it is background-dependent, meaning that the theory assumes a pre-existing kind of spacetime within which the objects of the theory exist. It should be noted that our Alternative Model is immune from such criticism.

Finally, there seems to be a cultural tradition among stringers associated with possible meanings of the "M" in the theory, suggestions including "Matrix", "Mystery", even "Mother", etc. To these it might not be inappropriate to add the name "Möbius"!

35

Another sine–Gordon System

Sine–Gordon solitonic behavior is of course not limited to our toroidal model, the definitive example being simply the pendulum of classical dynamics as discussed in Sec. V. Another intriguing example is the ***Phase-lock loop***, a feedback device ubiquitous in the electronic implementation of signal processing systems such as radio, radar, television, etc., for which it is necessary to maintain synchronism between received carrier signals and local references. Such systems are prone to chaotic behavior but their ability to initiate and maintain lock is well known and is herewith anticipated to have bearing on the genesis and stability of the particles of our model. Figure 35-1 shows a prototypical analog phase lock loop (PLL) with input and output $V_0 = A\cos\phi_1$ and $V_4 = \cos\phi_4$, respectively. Although the actual implementation is somewhat more involved, as per the feedback path, the two signals are basically multiplied together to generate a signal that encounters the several indicated operations. Thus, LIM is a Limiter that eliminates the input amplitude A. LPF is a minimal type of low-pass RC filter as shown in Figure 35-2 (R and C standing for resistor and

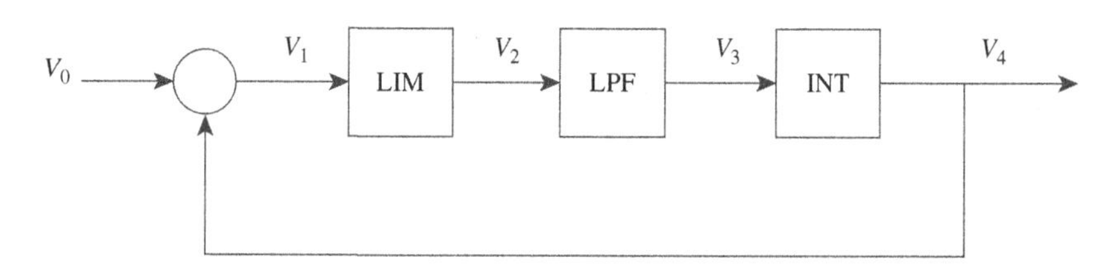

Figure 35-1. Phase lock loop schematic.

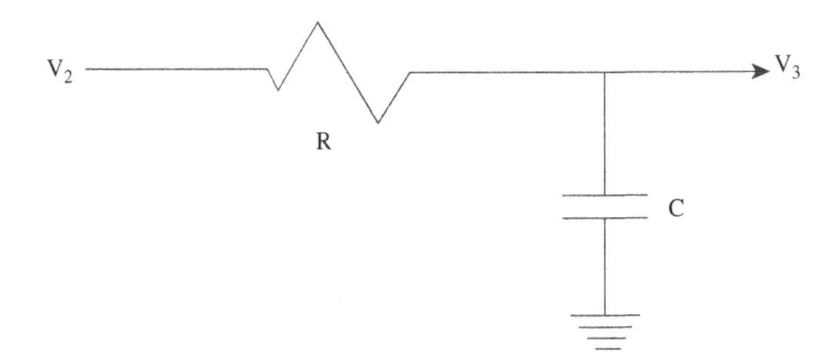

Figure 35-2. Basic analog low-pass filter.

capacitor, respectively) with transfer function in Laplace notation $V_3 = V_2/(1 + \tau s)$, where $\tau = \mathrm{RC}$.

As an operator in real time, variable s translates as a differentiator. Finally VCO is a voltage controlled oscillator whose transfer function is 1/s in Laplace notation which represents an implicit integration, implicit because while its ouput is the ***frequency*** of a sinusoid, it is its ***phase*** that is compared with the phase of the input. For large enough values of RC the LPF is also an integrator with transfer function, $1/\tau s$. In effect, the overall transfer function of the PLL is therefore

$$V_4(s) = \left(\frac{1}{\tau s^2}\right) V_1(s) \tag{35-1}$$

or

$$V_1(s) = \tau s^2 V_4(s) \tag{35-2}$$

which translates to

$$\frac{d^2}{dt^3} \sin\phi_2 = \left(\frac{1}{\tau}\right) \sin(\phi_1 - \phi_2) \tag{35-3}$$

or, after a bit of algebra,

$$\left\{\ddot{\phi}_2 + \left(\frac{1}{\tau'}\right) \sin\phi_2\right\} = \left(\frac{1}{\tau}\right) \sin\phi_1 + (\tan\phi_2) \sin\dot{\phi}_2, \tag{35-4}$$

which has the basic ***sine-Gordon*** *impulse response* on the L.H.S. and a forcing function (the first term) and damping (second) term on the R.H.S.

As to the kind of signals we are talking about, we recall the remark in Sec. V that the projection of a circular helix onto a plane including or parallel to the cylindrical axis is a sinusoid. Although the situation in the case of torus knots is rather more complex, the particles of our model can also be associated with a ***signal structure*** made manifest by projection. For example, the projection onto coordinate x was expressed in Sec. III in terms of a carrier signal at frequency, f_m, and a pair of sidebands at $f_+ = f_m + f_n$ and $f_- = f_m - f_n$, as

$$x = \frac{R\cos 2\pi f_m \ell + r}{2}(\cos f_+ \ell + \cos f_- \ell), \qquad (35\text{-}5)$$

where m/L is defined in terms of the traverse $\phi(\ell) = 2\pi\, m\ell/L$ such that $\phi(L) = 2\pi m$ and, it will be recalled, m and n are the number of longitudinal and meridianal traversals, respectively.

As noted in Sec. III, the sideband contribution disappears as r/R goes to zero but the basic topological nature of the MS (and the implicit knots it subsumes) ***cannot disappear***; the twist (or the implicit toroidal winding) is ***essential***. In any event if we assume that the rate of change of ϕ is the same for all particles we have a ***unique*** signal associated with ***each*** particle type. Figures 35-3 and 35-4 show the effect of varying the relative magnitudes of R and r for the case of $m = 2$, $n = 3$, the trefoil. The behavior in Figure 35-3 with $R = 0.95$ and $r = 0.05$ is due almost entirely to the carrier, $m = 2$ and we see two cycles of x carried out over the span of ϕ variation but for $R = 0.65$ and $r = 0.35$ as per Figure 35-4 we see the distortion due to the effect of the sidebands. We note that from another but related point of view, the XY plots may also be considered to constitute generalized *Lissajous* figures constructed from the periodic behavior of x and y versus ϕ.

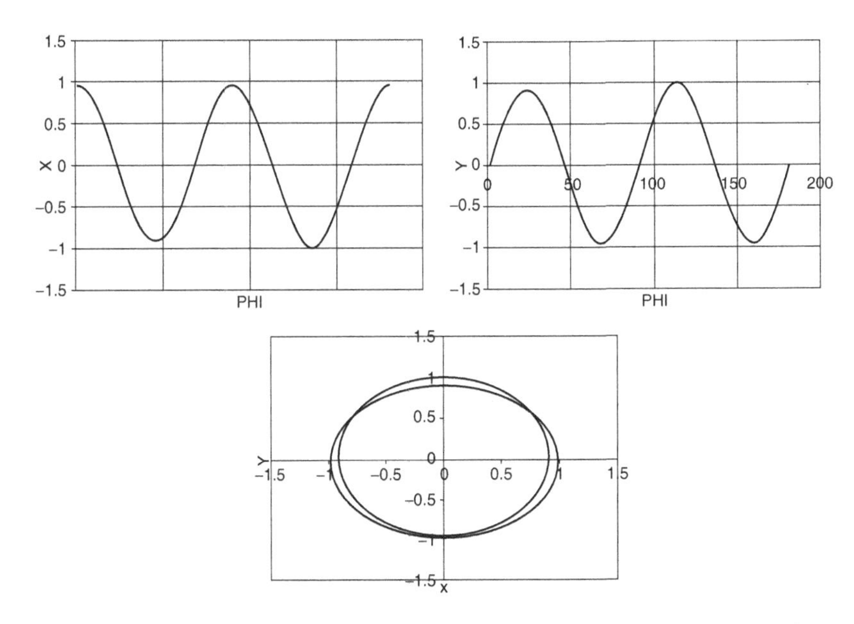

Figure 35-3. x and y signals for trefoil (small modulation: $n = 3$, $r/R = 0.05/0.95$).

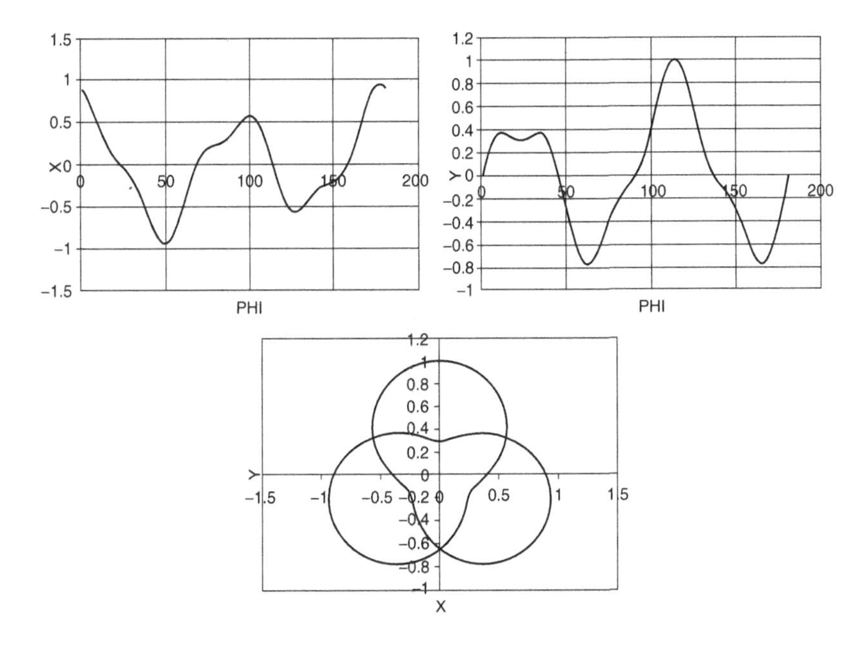

Figure 35-4. Same as Figure 35-3 but for large modulation; $r/R = 0.35/0.65$.

VII
Perspective

A Look Back and Forth, In and Out, Around and About[1]

[1]I must apologize for Chaps. 2 and 3 where we wander up and down the cosmic scale quite a bit as I subject the reader to considerable cosmological uncertainty. Sorry, but it's hard to be "down to earth" when the subject still seems to be so "up in the air"!

36

Recapitulation

The objective of this chapter is indeed to recapitulate and summarize what we have said about our alternative model of the elementary particles — its salient features, how they interrelate and their connections to historical antecedents and to the subject of elementary physics in general — in other words, basically to provide some perspective. In that regard it might be a good idea to resurrect the elementary question: "Just exactly what ***is*** an elementary particle?" Or, perhaps "what do we mean when we use the terminology, 'elementary particle'?" might be a better way to phrase the matter. In effect, the answer in this book has been and remains that we have arrived at particle elementarity when *there is* ***no need*** *to continue the search* for it by further process of reduction because the consequent relationships and phenomena appear to be those we want!

Which sounds tautological but that is alright; it brings us back to the model, and as we have seen, this is a model for which the process of reduction ***stops*** at the level of nucleons and electrons because particle attributes and conformation in space are ***linked***, inherently and inseparably at that level. Thus, it would appear, there is, indeed, no need for further search for such elementarity.[1] At this point, we can do no better in summing up the model's approach than to quote the following from the introduction to the first paper published in a knot theory journal dealing with

[1]Not to say that there are no more questions that need answers; far from it and I am sure you have a bunch. I will talk a bit about such things later on; there is still lots of work to do.

this model [10]: "— Here, beginning with two rudimentary knots, the unknot and the trefoil knot, we develop a unique approach to understanding the elementary particles of physics in terms of a visualizable reduction of all particles — fermions and bosons, hadrons and leptons — to a common topology — particles are regarded not as discrete, pointlike objects in a vacuum but as *localized distortions in and of an otherwise featureless continuum that supports torsion as well as curvature.*" We should also reiterate (See Sec. II) that these two knots are the simplest versions of (2, n) torus knots which, in turn, are equivalent to two-strand braids with closure, both, especially in their extension to Möbius strips, being the most elementary way to realize the 2 to 1 coverage of the rotation group SO(3) by the gauge group SU(2).

Several points in the quotation are noteworthy: one, of course, is the "Sakata-like" nature of the model, the fact that all particles belong to the same genus and, as per the above, follow from the "elementarity" of two rudimentary members of the same genus, the trefoil and the (folded-over) unknot. It is also important that the model is ***visualizable*** in which regard it is much like the diagrammatic elaborations of chemistry. (It is noteworthy that, conversely, chemistry is increasingly finding use for knot theoretical expression and investigation [27].) In that regard, this might be a

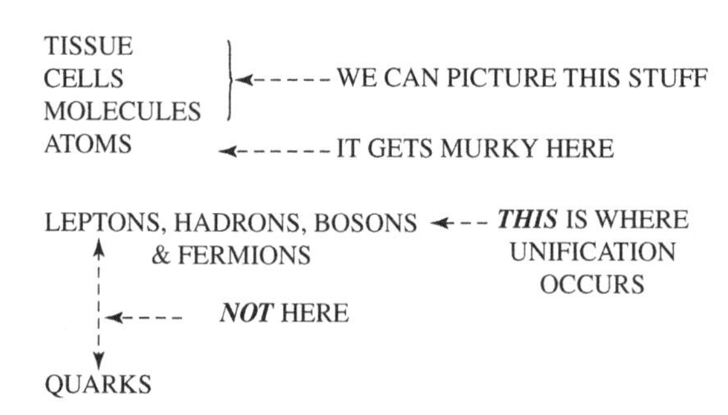

Figure 36-1. Overview (repeat).

good time to repeat the figure shown in the Introduction, the one that summarizes that aspect of the model as well as the way in which it halts the process of reduction.

Clearly, there is no multiplicity of fields in our Alternative Model; the continuum of reference here is just ***spacetime itself***, a situation that is the first, but not by any means, the least important hallmark of the model, its ***austerity***. In fact, there really is quite a small set of basic notions, an axiomatic basis so to speak, for the model. They are:

- *Toroidal Topology*: Möbius strips or their knot theoretic equivalents.
- *Twist*: It appears automatically but it is still a fundamental invariant.
- *Flattening*: Trading twist for localized writhing.
- *Traverse direction*: It must be specified.
- *Time*: As the odd dimension in a $2 + 1$ dimensional format.

From the point of view of ***differential*** geometry, it is the ***toroidal*** topology of a knot wound around its putative toroidal last that gives us its identity in spacetime as a ***soliton*** as well as that of a nontrivial vector ***bundle***. And from the point of view of ***algebraic*** geometry, that topology manifests the pairing of two group identities; locally, it looks like U(1) with a connection that implies a field of some kind and, overall, an associated charge. At the same time, its ***twist*** is the source of both ***isospin***, algebraically additive in combination, and ***spin***, additive in combination in multiples of 1/2, thus implicating SU(2). As recounted in Sec. I, it was the sight of a ***flattened*** MS that led to the small set of basic fermions ultimately recognized as equivalent to the **Sakata** approach to model building. Flattening also gives us the two-dimensional FMS ***planform***, and, in concert with the choice of direction of traverse, the two kinds of quirk that end up with the ***fractional charge*** of the SM's quark model. But without a ***traverse*** direction, our small basic set exhibits complete mirror symmetry; it is the ***choice*** of traverse that ***breaks the symmetry*** and solidifies the actuality of ***fractional*** charge. And it is the ***reversal*** of traverse direction that defines charge conjugation and ***antiparticles***.

Finally, identifying the out-of plane dimension with ***time*** gives us CPT invariance and the notion of antiparticles as particles moving backward in time. Once this "axiomatic" basis is in place, there follows, quite automatically, the development of a set of interaction models and a taxonomy organized in terms of the product of SU(2) vector ***spin spaces***, followed by the combinatorial analysis of degeneracy in terms of ***twist*** and ***charge*** categories. The formalization of combinatorial analysis as introduced in [12] is implemented, we recall, in terms of symbolic ***convolution***, and the analysis of detailed FMS composition and contingency in terms of the inner product of symbolic ***vectors***.

Not surprisingly, in keeping with the thematic emphasis on it in the preface, a noteworthy aspect of the development is a ubiquitous, ***geometrical*** element that encapsulates the model's taxonomy. In Sec. II we saw this graphically in the way the three-dimensional edifice of second order fusion is organized in terms of inclined ***twist planes***. This is mainly an ***algebraic*** geometrical consequence of the model's toroidal topology. Additional such algebraic geometrical considerations are displayed in the ***complex*** algebra and the ***quaternionic*** algebra of Sec. III. However, as we have seen, the topology also constrains the particle's ***differential geometry*** and its relationship to General Relativity in a way that allows its solitonic nature to emerge and, in fact, even to validate our choice of traverse direction as the ***determinant*** of particle/antiparticle identity.

So much for the basics of our model. In terms of their expansion into a taxonomy, a basic feature is the notion of ***bound*** states of fermions and antifermions, created in the model by fusion and here again we must cite the anticipatory publication of ***Fermi*** and ***Yang***, discussed in Sec. V, specifically with regard to the charged pions featured in that section. As emphasized by **Nambu**, this was also the point of departure for ***Sakata*** in his attempt to enlarge the prevailing mid-20th century taxonomy to accommodate the emergence of "strange" particles. On the subject of bound states, we also recall the manner in which the ***Dirac*** equation yields, as per the treatment in Sec. II, a not-generally-emphsized result, namely the ***interdependent*** existence of a particle and an antiparticle in what can be interpreted as a bound state (and, in a way, implicates the Alternative Model's connection to Quantum Mechanics and Special Relativity).

Möbius strips and torus knots are also exemplary in their ***inter-dependence***, whether we describe the latter as the boundary of the former or, conversely, the former as the concatenation of the latter. In terms of the knot connection to particle physics, we recall once again the 1917 article by ***Einstein*** in which he essentially defined what amounts to a torus knot as a way to ***unambiguously*** describe the quantized trajectory of a particle in a central field of force thus, in a sense, also anticipating our model. And, of course, never-to-be-forgotten are the valiant mid-19th century efforts of ***Kelvin*** to model atoms as knots in the "ether". The Einstein and Kelvin contributions were isolated cases and not of lasting consequence (except, perhaps, as resuscitated herein!). Nowadays, however there is a large and growing literature on the connection between knots and physics [26, 28], not the least of which involves the connection between knots and topological quantum field theory (***TQFT***) as discussed briefly in Sec. VI. Another connection to physics and in fact, to mathematics as well, is the relationship of the alternative model to the subject of Quantum groups/***Hopf algebra*** as discussed in some detail in Sec. III. In both the TQFT and Hopf cases, specific requirements put the operations of fusion and fission under detailed scrutiny. In the Hopf case a further requirement (the existence of an "***antipode***" [33]) turns out to be satisfied in a unique way specific to the model. In either case the conclusion is that the requirements are satisfied. Some of the connections to String Theory are also discussed briefly in Sec. VI. It is quite possible that more such connectivity will emerge as time goes on, depending in part on the extent of the readership experienced by this book. I am open to all suggestions!

Section V also compares how our Alternative Model and the Standard Model generate particle ***mass***: in order to implement the symmetry-breaking so vital to its basic theory in that regard, the SM ***imports*** a field and the associated particle, the **Higgs** boson. On the other hand, the AM embodies mass automatically by virtue of the underlying particle ***topology*** plus a local variation in the parameters usually associated with gravitation, a variation therein associated with a combination of that 1917 Einstein concept of mapping the orbits around a central field of force onto a torus and Niels Bohr's atomic theory adapted to gravitational attraction. Also, symmetry breaking in the AM is manifested in a dynamic way as the particle's local solitonic distortion circulates longitudinally around its toroidal "last", all longitudes being equally likely, a circumstance sometimes invoked in discussions of Goldstone's Theorem [49].

Also, you may recall, early in the course of the development of that internal gravitational variation, a serendipitous connection to the so-called "strong" nuclear force emerged thereby unifying seemingly disparate phenomena and leading to a bit of introspection: in retrospect, it would appear that the labels "strong" and "weak" as applied to particle interactions are misleading misnomers; in the AM there is only the ***one*** coupling constant for ***very short*** distances; it is associated with a "strong" interaction between quirks and antiquirks and takes place in very short times. "Weak" interactions on the other hand are really so labeled, in my understanding, because they take place over longer times, something that is indeed quite understandable since, as per the modeling of Sec. IV, more and more complex things have to happen in a weak interaction including coupling and/or its converse "uncoupling" (Fusion and Fission herein).

Of course, the origin of such local variation as remains and its precise, repetitive association with each member of a given type of elementary particle evokes fundamental questions which, when viewed in a wider context, are not really new; questions as to why certain particle types and no others exist (call it the "exclusivity" problem) and why the members of each type are identical (call this the "identity" problem) have, heretofore,[2] never been satisfactorily answered except, perhaps, from a pair of diametrically opposed points of view. One is couched in so-called "***Anthropic***" terms for which no universal agreement exists as to which of a number of versions to espouse, but where matters of ***belief*** inevitably emerge

[2]In a way, our model solves the exclusivity problem by definition!

for all. The other is also a "multinotion" notion, known collectively as the "***Multiverse***"; it postulates the never-ending creation of universes possibly differing from each other in their basic physical laws or at least parameter values. The first viewpoint has (in some versions) Mankind at the "focus of Creation", in fact, as the "Raison d'etre — the reason for something rather than nothing! The other diametrically opposite viewpoint makes the question statistically irrelevant because, given opportunities without limit, we were bound to show up sooner or later! We might call that the ultimate "Deflationary theory![3]

And, while I certainly have no quarrel with Quantum Field Theory — it is a marvelous mathematical system (in fact the system of sine-Gordon solitons are generally regarded as realizations of QFT although that has not been emphasized in this book) — its answer to those questions, namely that each type of elementary particle is just the quantum of a particular field, would appear to beg the question since the mystery, the way I see it, is thereby just transferred to the associated array of fields.

[3]We would just have to get used to the idea: we cannot get any less important than that!

37

Cosmology, Beta Decay and Anthropic Matters

Remember the chart we showed at the end of the Introduction? Here it is again (Figure 37-1) and at this point we are more or less at its right-hand side (at the top in this display!).

We have talked about dark matter and TQFT and there is that big block labeled TOE. But, no; you can sit back and relax; I do not anticipate coming up with a "Theory of Everything" any time soon! Actually, I do not think anybody will, either sooner or later; I just stuck it in there because people like to talk about it as sort of a natural culmination of all the good things that have been, or should be coming out of more and better experimentation, measurement and theorizing. Nevertheless, it looks like we may be able to make some connections to matters of cosmic import and that is what this chapter and those that follow it are about.

We ended the last chapter with a disconcerting gap in our vast/2 store of knowledge. With apologies to William Shakespeare, we might characterize what's missing as, to paraphrase Juliet's plaintive outcry:

"Particles, particles, wherefore art thou particles and thine divers varieties, and no others, with each such per variety all of a sameness?"

That is supposed to be sort of the way Will might have had his people talk about our subject matter in those days (I make no claim to expertise in this area). But, actually, can you imagine what the people of that time might think of the kind of stuff we are bandying about with such abandon nowadays? "Sorcery! A witch's brew!"

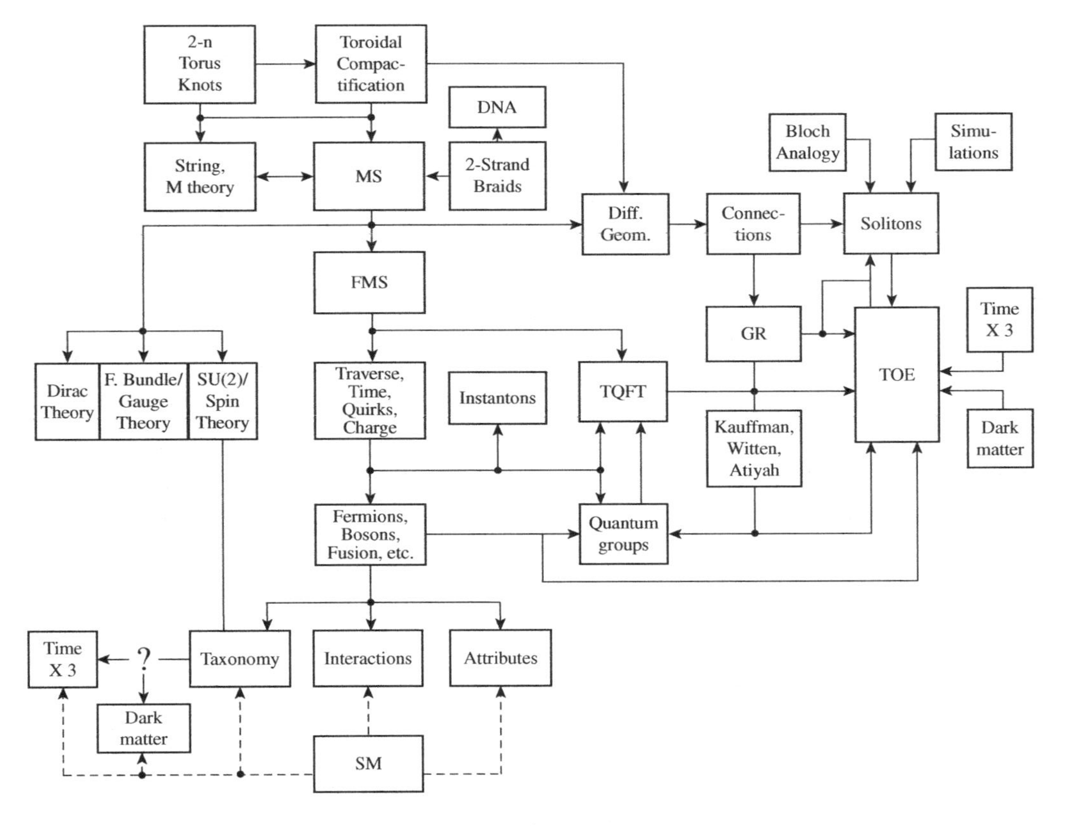

Figure 37-1. Connections (repeat).

Which brings us back to current matters of belief and the Anthropic point of view mentioned above. I do not think there is anything mystical or nonscientific about the notion that our universe must be such as to harbor our kind of life because here we are talking about it! Some of our best-known and productive physicists have investigated such matters in all seriousness. And as for the "Multiverse" idea, while it may seem hard for many to get used to, as of this writing there may be a Nobel Prize in it for at least one cosmologist![1]

Personally, I suspect that answers to such questions might be lurking in the ***Machian*** notion that the entire universe is in some sense responsible for local behavior ***if*** coupled with the (admittedly tautological) notion that it, itself, is simply the sum total of its component parts! We would then expect to discover a spectrum of behavior characterized by ***eigenstates*** in some way, shape or form as a result of the implied ***closure***, which, among other things, might actually imply a never-ending, repetitive emergence of universes, possibly with that statistical behavior alluded to above.

Many years ago — make that many decades ago — when I took my first physics class in college, although I had not yet heard of Mach's principle, I had a similar, but ill-formed notion; namely that the *entire universe was,* ***somehow****, contained within each of its constituent parts*! In logical terms that is sort of a ***self-referential*** notion, but it is ***not*** exactly the kind of thing where closure occurs between terminating elements of ***similar scale***, something we expect to see in homeostatic or feedback control systems, (or, more generally, in self-referential, second order differential equations). No, what we are talking about here are connections between the super-macroscopically ***large*** and the sub-microscopically ***small***. And, what, you might ask, was the reason for invoking such an outlandish notion? Well, as I recall, I was looking for a universal, unchanging entity of ***reference*** and the only one I could think of at that time was the universe itself!

[1]Andrei Lindei has been mentioned in the news probably also for his connection to Inflationary theory.

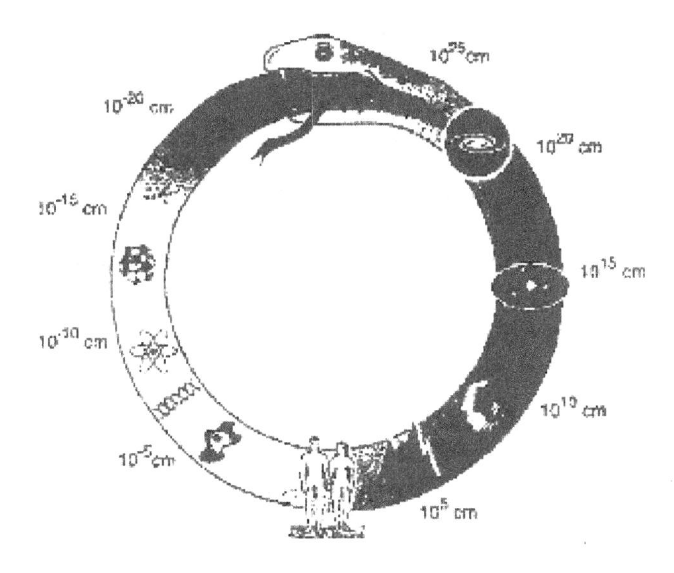

Figure 37-2. Ouratio the Ouraborus.

I will return to that youthful notion (YN) later on but in the meantime, perhaps you are familiar with the ancient imagery of the ***Ouraborus***, the snake that swallows its tail, with closure ***somehow*** occurring between ***maximum*** and ***minimum*** scales as we see at the top of Figure 37-2. Although the connection suggested by that herpetological imagery (You may have learned a new word here!) does not quite apply to my YN (its inverse to it, right?) it is such an evocative idea and of such ancient origin that we ought to talk about it a little bit. First, I call your attention to the numbered pictures along the creature's corpus indicating ***scale*** (and the little figures on the bottom — they represent us — mankind!); they are of course a modern addition; you may have seen other depictions with other markings. And, in a way we are back to the discussion you may recall, early-on in the Preface about looking up, out and around as well as down and in: In other words, starting at the bottom and proceeding counterclockwise on the right of the picture, we contemplate a larger and larger scale until, eventually, we are confronted with the ***Universe*** itself! Conversely, going up clockwise from the bottom we see a smaller and smaller scale until we are faced with "***elementarity***". Actually, if we start at the top and proceed clockwise all the way around, we are in a completely ***reductionist*** mode, cycling down through about 60 orders of magnitude!

Its ancient origins notwithstanding, the Ouraborus often emerges (just like here!) in modern discussions of the possible relationship between the cosmos and the elements of which it is constituted. For example — and here we are going to take a short trip into a particular issue of interest to our Alternative Model — the authoritative astronomer **Martin Rees**, in a book that shows that exact picture right up front [62], lists ***six numbers*** whose empirical values (not yet amenable to known methods of prediction) are, within narrow limits, necessary for our universe, ***and*** the life it supports to exist in the way we experience them. Of particular interest in terms of the deliberations of Sections V, above, is that Rees also mentions a ***seventh*** number, the ***ratio*** of ***proton*** and ***electron*** masses as an example of something of universal recognition that might expedite communication with extraterrestrial beings were such found to exist. Which, of course, naturally brings up the question as to whether this number, also, ***must*** have the value we measure for it.

So, it is noteworthy that, in a monumental analysis of the ***Anthropic Principle*** and its various shades of interpretation, **Barrow** and **Tipler** [63] examine the relationship between cosmological and elementary particle physics in considerable detail, ***stressing*** the importance that the mass of the proton (938.272 MeV) be less than that of the Neutron (939.566), especially in terms of the weak interactions. The authors point out that the difference between the two masses ($\Delta m_N = 1.293\,\text{MeV}$) is roughly on the same order of magnitude as the mass of the electron (0.511 MeV) but ***larger*** by an amount usually denoted by $Q = 0.782\,\text{MeV}$, the so-called "energy yield" of the neutron's (beta) decay into a proton, an electron and the electron's antineutrino.

You may recall the discussion of Sec. V of the relationship between a particle's mass and its size. Among other things it led to a hyperbolic kind of ***constraint*** between the ***size*** of a nucleon and that of the electron; when ***one goes up***, the ***other must go down***. In the context of our beta decay model, that may be more important than its mass per se (we repeat the associated diagram here in abbreviated form for reference. Note the crossover point, X).

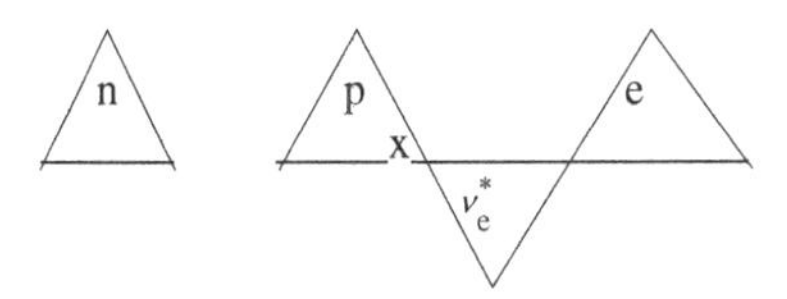

Figure 37-3. Beta decay schematic.

To explore the ramifications of incremental size a bit we recall from Sec. V the relationship between mass and size which in terms of the nuclei we can write simply, as $r_n/r_p = m_n^2/m_p^2$. It then follows directly that the ratio of the size ***decrement***, $\Delta r_n = r_n - r_p$ between the two nucleons to the size of a nucleon as per our model is expressible as

$$\frac{\Delta r_n}{r_n} \simeq \frac{2\Delta m_n}{m_n},$$

which evaluates to $\Delta r_n/r_n \approx 2 \times 10^{-3}$, about four orders of magnitude larger than the relative size, $r_e/r_n \approx 2 \times 10^{-7}$, estimated for the electron in Sec. V. In other words, the linear space available to the electron on the right-hand side of the beta decay diagram, after the proton takes up its share, is much, much more than it needs.

Thus, although exactly where the original folding and ultimate breakup of the neutron occurs is not predictable at this point, apparently (what becomes) the electron requires a great amount of size ***reduction***, percentage-wise in order to finally qualify as a bona fide electron while, correspondingly, the nascent proton would need to ***expand*** but only to a much, much lesser degree. Thus the above ***constraint*** operates here to ensure that the motion of point x in the figure is in the right direction. Also, qualitatively speaking, the discussion of curvature versus torsion as presented in the differential geometry of Sec. V, indicates that the electron is more prone to ***curvature*** than the proton and, conversely, the proton is more prone to ***torsion*** than the electron, implying that, given some region in which to expand or contract for each, what might be termed the "proto-electron" would, indeed, tend to end up much smaller and the "proto-proton" would tend to end up "somewhat" larger.

Of course, all the processes involved in this model of neutron decay are inherently statistical which implies a thermodynamic overview and, in a sense, we have been discussing what amounts to a generalized kind of thermodynamic phase transition. Which gives us a rather sneaky segue into a little different kind of phase although not, initially, in the thermodynamic sense; that of a periodic process. I'm referring to the ***solitonic*** modulation induced upon spacetime as discussed in Sec. VI, with a carrier and sidebands characteristic of the size of each particle. Since, as discussed above, the decay products of neutron decay must ***evolve*** into their final characteristic size, we should expect a ***transition*** history into a final phase-lock condition for each such modulation as particle size increases or decreases to its nominal value, thus completing the decay process. And, as in other phase-lock processes, we should expect this one to also proceed in a statistical manner with an associated thermodynamic overview.

At the same time, it is incumbent upon us to keep in mind that the R.H.S. of the constraint relation mentioned above (The "Interjection" in Chap. 28 of Sec. V) involves the speed of light, c, Planck's constant h and the gravitational "constant", G; we will have more to say about that in Chap. 41, quantum gravity.

Well, at this point, although it looks like something that ought to be explored further, that's about all we have to say in this final foray into Beta Decay. Right now, let us get back to the cosmos and the possibility of its potential eigenstate behavior we started to talk about in the preceding.

38

Herpetological Closure

Our cosmic snake with the enormous appetite is not the only way to portray interdimensional closure. In fact, as mentioned above, it is not exactly the kind of closure I had in mind all that long time ago. Another possibility is in terms of the so-called Klein Bottle, a sort of topological generalization of a Möbius strip (in fact slicing it the right way yields a pair of Möbius strips, with $\text{NHT} = \pm 1$). It features closure but in four dimensions and, in fact, in three-dimensional depictions, an extension of the entire bottle is inserted into a hole cut into the bottle itself, a kind of surgery not necessary in the actual re-entrant phenomenology of four dimensions. In any event, we are not prepared to talk about that now; maybe next time.

Something that appears more promising in terms of closure was developed by Louis **Kauffman** in a fascinating paper [64] on matters of reflexivity, recursion and so forth, and directed toward the bases for some of the fundamental notions of physics and mathematics. What follows is abstracted from a communication [65] of the underlying considerations intended for ultimate publication. Kauffman, also makes reference to the Ouraborus but he then proceeds to take a different approach, examining in minute algebraic, logical and pictorial detail how ***reflexive*** behavior can produce what he calls "***Eigenforms***" where reflexivity refers to the presence of a relationship between an entity and itself. His primary interest he defines as a reflexive ***domain***, an example of which is "A market or a system of finance is composed of actions and individuals and the actions of those individuals influence the market just as the global information from the market influences the actions of the individuals." That is what really caught my attention because it is reminiscent of my youthful notion; that, you may recall, in some way shape or form, the entire universe is contained within each of its component entities!

Kauffman's ultimate goal is to show that eigenforms, that is, "— ***fixed points*** of transformations, are present for all transforms of that reflexive domain." In that regard a noteworthy remark is "The notion of an eigenform is inextricably linked with second order ***cybernetics***. One starts on the road to such a concept as soon as one begins to consider a ***pattern of patterns***, the form of a form, the cybernetics of cybernetics. Such appear to loop around themselves —" (Emphasis added). To which we might add that cybernetics may be interpreted, as per the above, to include second order differential equations and feedback control systems such as phase lock loops! And even sine–Gordon solitons!

Kauffman uses a progression of (deceptively simple!) pictures to illustrate his arguments: First there is an unspecified entity X and an operation, $F(X)$, on it that puts it in a box. The picture employed is a two-dimensional representation showing a ***square*** with an X inside it but, at least for our purpose in what follows, it is to be understood that the sides of the box do not really exist; the box is a metaphor for a ***set*** or category, in fact one of not-necessarily-defined generality at this point. Nevertheless, to continue, if X, itself, is a box containing such an entity, one can set up a ***recursion*** featuring a ***nested sequence*** of boxes, such that, eventually, the process becomes essentially invariant to the addition of just one more box and in the limit, the result can be expressed as $F(X) = X$, or $X = F(X)$. In fact, we note, that this process of ***recursion*** is actually ***self-referential***, which is expressed ***symbolically*** in the paper as a ***feedback*** process for which X is an ***eigenform*** for the process $F(X)$. Or, substituting $F(X)$ for X in $X = F(X)$ gives $X = F(F(F(F(X)))) = \cdots$ which, according to Kauffman is a way of showing that the eigenform, X, is what he calls "an implicate order for the process that generates it".

Now, Kauffman goes on to expand upon his procedure, using it to form the conceptual bases of complex arithmetic, exponential functions and Quantum Mechanics! However, he is quite clear in characterizing it as a strictly theoretical process, something that ***exists*** only in the ***mind***; as we stressed above, the boxes are not "real"; they are certainly not tangible nor do they have well-defined boundaries. Instead, they are just ways of thinking about the process of inclusion. Of course, I agree. But, what, if anything does all this have to do with cosmic snakes swallowing their tails?

Well, in what follows we shall explore the application of the box procedure to our universe and its history, keeping in mind that what we will be doing is figurative only, a conceptual exercise and, harking back to the message of our Preface, an allegorical enterprise. Given those qualifications, it would appear that what the box sequence procedure provides is an ***alternative*** to, but equivalent, and actually more appropriate, point of view than the Ouraborus (which, of course, is, inherently allegorical). You may recall the unspoken question I posed when I introduced our herpetological friend, Ouratio, up above; what I said is that closure between maximum and minimum scales occurs ***somehow*** but in a way not really suited to the problem I wanted to talk about, namely the way that the universe influences elementary particles. Nevertheless, the point here is that the (self-referential) box sequence procedure promises to provide a feedback path that ***does*** *implement* that closure in such a way.

So, let us see how that might work as applied to our universe: for example, suppose we start out by defining X in the above as the ***set*** of basic elementary particles of our experience with subsets being the electron the proton and the neutron. The next step is to identify a "box" to put X in and for that, suppose further that we just sort of cycle up through the kinds of things we talked about in the Preface; first the set of ***all*** particles including the taxonomy formed by the elementary particles — that's box #1. Box #2 is then the set of ***atoms*** (meaning all those in the Periodic table of the elements rather than just those associated with terrestrial life). Then, since it's made up of those atoms, the Earth, itself, followed by the solar system, and similarly, on up to our galaxy, the Milky Way, then the neighborhood it inhabits, and on up through all the galactic patterns, the patterns of patterns, etc., etc. and, finally — and here it gets a little murky — to the extent that we can sensibly talk about it, our Universe itself. Then, assuming we can associate a size to each such "box" and if we compress their labeling, perhaps using a logarithmic scale, or some such — we need not belabor the point — eventually the numbers do not change that much, percentage-wise, from one step to another, thus endowing additional legitimacy to the use of Kauffman's results.

But wait a minute: something strange — paradoxical might be a better word — is going on here. If we keep that up, trying to imagine boxes that encompass entities of larger and larger ***scale***, then, since larger scale means longer ***distance*** from Earth, we're also going back in ***time***. Our expertise of reference, here, is that of the Cosmologists, and according to the Standard Cosmological Model, our pursuit of ***larger*** scales is actually taking us ***backwards*** down the history of the "inflationary"[1] expansion from the so-called Big Bang, an essential element of the SCM, down to ***smaller*** scales! And at some scale, we are going to be looking at times before there even were galaxies. Or even stars! And since most of the elements in the periodic table are understood to be cooked up in stars, we will be down to just Hydrogen, Helium and Lithium, and, in fact, eventually, perhaps just elementary entities as they emerged from the Big Bang. In fact, all the way down toward that primeval time before there was elementary anything! That is scary; I do not even want to talk about going backwards through that last phase, the inverse of Inflation, call it ***deflation***, toward Oblivion!

So let us just stop this helter-skelter recursion for a moment and take stock: let us stop at that point way back in time when those elementary entities were putatively just emerging from the

[1]Many if not most readers are probably aware of, if not conversant with, the idea of the humongous inflationary period of expansion of the (or each) universe following its initial manifestation.

Big Bang. In terms of space and time we are down to times maybe 10^{35} shorter than a second after the inflationary epoch and a universe that is something on the order of a millimeter in size! The whole universe! Can you believe it? But it gets worse; it gets down to sub, sub, etc., microscopic sizes! We're now deep down in the realm of particle physics, appealing to the expertise of particle physicists, much of whose knowledge and intuition in this regard comes from extrapolation from the Standard Model of particle physics and high energy experimentation at CERN and Fermilab.

And what they say is that the elementary entities we encounter are just the ***same exact ones*** we have today, the ones they deal with. *Well,* ***of course****, they are*! Consider: Starting out with a set of elementary particles, familiar from terrestrial particle physics, and pursuing larger and larger scales counterclockwise along (or within?) Ouratio's corpus, we have finally encountered a set of primeval particles that particle physics theory tells us, lead (in real time) to the taxonomy we have today! ***Current particles*** meet **primeval particles**! Look familiar? It ought to! It is kind of like you take a trip in a time machine and you meet your great, great, etc. grandfather; he looks like you, he talks like you and if you poke him in the ribs it makes you wince!

So, does this mean that the box procedure worked or not? And that the elementary particles of which everything is currently constituted are, in fact the "Eigenforms", (to use Kauffman's nomenclature) of our universe, our box of all boxes? Or not? Well, in the first place, as per Kauffman's admonition, the journey we took was strictly conceptual and there are indeed problems with taking his procedure out of ***context***; we will talk about that in a minute. As far as the box procedure itself is concerned, there is nothing wrong with it; it is fine. However, there is a problem with those particles the particle physicists believe emerged as the primeval universe cooled. And, as it says a paragraph or so back, that belief originates as a result of "extrapolation from the Standard Model and high energy experimentation at CERN and Fermilab." Which, of course uses the precise particles that exist here on Earth. In other words, at the end of our boxing expedition we are not just comparing apples to apples but a particular set of apples to ***itself***! In retrospect, we should have known that to begin with.

Actually we did; I mean I did and you probably did too. I rather enjoyed the trip and I believe it was worthwhile. But apparently we are back to square one; Juliet's plaintive query has still not been answered. If we were looking for answers to those "exclusivity" and "identity" problems we talked about above, it would seem that our recursion procedure has not really worked out in exactly the way it seemed it might when we began. But here I have to apologize; we did take the box procedure out of context! Nevertheless, it did its job; it looks like our elementary particles are indeed eigenforms of our Universe. But that is all the information the box procedure can provide; that is what it is designed to do. So, where does that leave us? Well, do you recall that quotation that appeared at the beginning of the book? It might not be too inappropriate to take another look at it so here it is:

> *We have found a strange footprint on the shores of the unknown. We have devised profound theories, one after another to account for its origin. At last, we have succeeded in reconstructing the creature that made the footprint. And lo! It is our own.*
>
> (Sir Arthur Eddington)

Ah, yes. So, having been thus appropriately chastened, let us then return to our exercise in boxing with a more discriminating point of view. Does my long-held notion of the universe's influence on elementary particles have to be consigned to the "dust bin of history", as they say, along with all those other puerile conceits that dwindle away as we grow up? Well, maybe not. Here is what I think we have to do to salvage it: First, with apologies to all the snake enthusiasts of the world, we need to perform an operation on Ouratio, himself. In order to modernize the way he looks he can no longer be allowed to swallow his tail because it is not quite compatible with the Inflation we described above as being, also, part of the SCM. So, as a result of the operation, Ouratio may no longer be allowed to have a head!

Which would imply some kind of radical procedure, but it has to be done; the image of the snake swallowing its tail is a fallacious image. The true image is a sort of herpetological ***loop*** with an obscure mating of what was previously the head and what is still the sub, sub, etc. microscopic tail. I do not know how to draw it because what was formerly the swallowing region is undrawable! It cannot be portrayed on paper because it represents the ultimate ***paradox***: the merger of the container with the thing contained; the mating of the very large and the very small; the very "raisson d'etre" of something rather than nothing; and the crucible from whence springs "Elementarity" and ultimately the presence of our kind of being. And apparently a descent into (thankfully) short-term lyricism.

But now that we have such a loop, we ought to be talking about its dynamics. In fact, we should be talking about the influence of a rapidly, actually violently, changing inflationary universe on the particle creation process, a universe that, according to both the cosmological and the particle people, is growing, cooling, and ***in the large***, ***flattening*** at a mind-boggling rate. (Incidentally, in a way, the very fact that we need the expertise of both cosmologists and particle physicists accentuates the mutual influence of the very ***large*** and the very ***small*** on each other!) And, although I don't subscribe to all of it (I guess you know my views of quarks), the particle people say that the entire universe, way back there before nucleons, was just a "soup" of quarks and leptons (the latter meaning electrons and neutrinos) just waiting for a further decrease in ***temperature*** to allow nucleons to emerge.

When I mentioned flattening, above, as being "in the ***large***", I meant on a universal scale, whatever that is at any given time, and in the realm of the cosmologists who deal mostly with large-picture, general relativity and thermodynamics. According to the SCM's large-scale picture, universal temperature exerts a go-or-no-go influence on particle emergence; each particle has a characteristic upper limit on the temperature at which it can exist such that when the temperature drops below the limit, the particle can "pop into existence" although I am not sure that anybody really knows exactly how that happens!

Meanwhile, on a smaller scale, the ***detailed*** picture I see is that of all kinds of variations in ***curvature***, in fact a whole spectrum of ripples in the fabric of spacetime. That is a local phenomenon and I believe it is there that we will find the answers to those pesky questions (Remember "equivalence" and "identity"). You may recall how, in Sec. V where we were developing the Lagrangian for solitonic behavior we invoked a relationship between energy and curvature. We can rewrite it here as $R \propto T/A$, where from the Einstein equation, R is the Riemannian curvature scalar, T is the energy scalar and A is the local stress or linear energy density. You may also recall how we developed an "indigenous parallel" to the Higgs potential energy characteristic used in the Standard Model to generate mass for elementary particles. What we came up with was a similar "***hill and valley***" behavior but ***internal*** to our particles in terms of that stress energy situation such that curvature was ***positive*** on a hill and ***negative*** in the surrounding valley.

And, in that roiling, primeval spacetime we left a minute ago, would you not think we would have lots of that kind of topography? That is what I believe, maybe not all exactly like what we talked about in Sec. V, but no matter; if we can characterize it, maybe we can say something useful about elementary particle generation. Which brings up the ***second thing*** we have to do to salvage our goal (remember the first requirement we stated a few paragraphs back?); we have to supply a way to ***identify*** our elementary particles in universal terms. If we could talk to Clifford right now; I think he would be fascinated and very gratified that his vision of little solitonic "hillocks" moving around in-and-of space are actually being investigated. However, to talk about that cogently, we have to include that identification feature and for that we need to go on to the next couple of chapters. If you are getting tired of this endless succession of chapters and a lot of speculation don't give up; we are getting close to the end of the book. But there is one more speculative notion I just have to tell you about, something that I would characterize as a Homeostatic or maybe Cybernetic depiction of my "Youthful Notion" but in terms of the technology of a hundred years or so ago (well, sometimes it feels like that!) when I was involved with such matters.

39

The Universe and the Marketplace

Perhaps you remember what I said in the Introduction about what I did to earn my keep when I was gainfully employed all those many years ago. A lot of it was involved with dynamic systems, most often linear, whose behavior, in one application or another, was described by an impulse response function in the temporal (or spatial) domain or, equivalently, by its spectral behavior, that is the response to each of a spectrum of frequencies usually denoted by its ***transfer function*** in the patois of that period, basically the Laplace or Fourier transform of the sysem's response to an impulse. The phase lock loop I talked about above is one such system but applications can of course be simpler, such as a thermostat or, more generally, a homeostat or even something more complex such as a large system with multiple inputs and outputs.

What I am looking for here is something we might think about as the "Transfer Function" of ***the Universe***, the implication being that we regard the universe as the ***Ultimate Homeostat***! In other words, just as life here on Earth maintains itself in a homeostatic way, so it is with the Universe, at every scale and, ***in particular***, in this chapter, down to the elementary particle level. Perhaps you remember the chapter in Sec. III that described a convolutional model of another "Kauffmanian" creation, the Bracket Polynomial and my block diagram showing a process that implemented it for (2, n) torus knots. I like to think of that process as a ***machine*** for making that kind of knot and I have reproduced its schematic here:

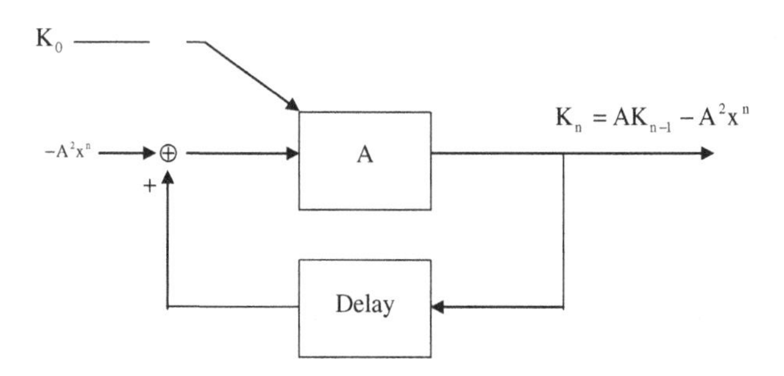

Note that the "identification" requirement we talked about in the previous chapter is satisfied by the parameter "n" which, you will recall, is the number of meridianal traversals completed in 2 longitudinal traversals of the putative toroidal "last" of the knot. With regard to variable A, you may also recall a comment in Sec. III to the effect that it emerged in connection with a well-known statistical process, the two-dimensional Ising model of a regular array of random up-or-down spins (Translated by Kauffman into under- or- over self-crossings of a knot in a two-dimensional array). In the present context, A is to be regarded in a more general way as an important dynamic variable, what might be viewed in circuit design as a complex system gain (amplification factor).

But now, suppose we consider our machine in a larger context so that it sends its output $K_n = AK_{n-1} - A^2x^n$ to another box labeled U, along with the outputs of ***many similar*** machines. Furthermore suppose that box U sends the letter A back to the box labeled A in each and every such machine. You guessed it; Box U represents the Universe and the machines represent our particles; they contribute to, and in return, get important input ***from*** the Universe that would be expected to govern their dynamic behavior.

We might view this situation as a particular ***manifestation*** of Kauffman's "market" a description I quoted above and repeat here: "A market or a system of finance is composed of actions and individuals and the actions of those individuals influence the market just as the global information from the market influences the actions of the individuals." To which, you may recall, I remarked that it reminded me of my youthful notion that the entire Universe is, in some way shape or form, contained within each of its component entities.

Here's what that enhanced system would look like, also in closed-loop, feedback block diagram form:

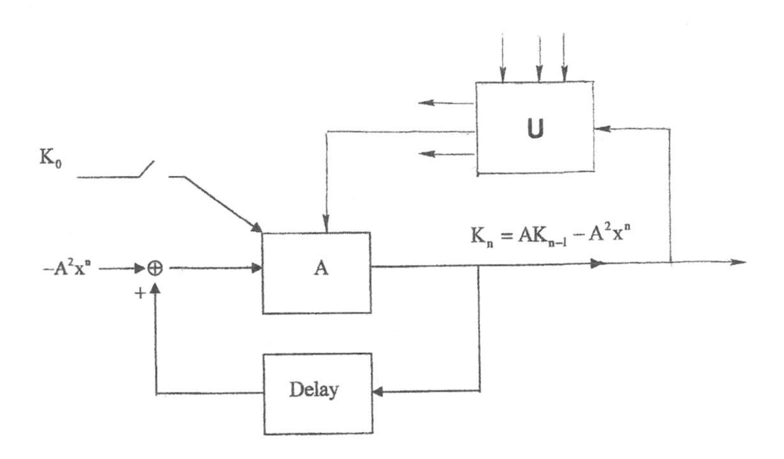

Figure 39-1. The ultimate homeostat.

What we would expect the "transfer function" of such a self-referential system to portray, in particular is the dynamic characteristic frequencies in ***time*** and wave numbers in ***space*** in the neighborhood of the smallest distortions of spacetime — Clifford's "ripples" — we would expect to at least ***include*** the inner Higgs-like structure of our elementary particles. Of course, that box labeled *U* representing our universe might be quite complicated as might the universal message including the "gain", A, it broadcasts to its elementary constituents so that the overall spectrum might continue on up to larger scales eventually to include the stellar systems, the galaxies galactic clusters, superclusters, etc. Similarly, the temporal behavior might include a large variety of characteristic behaviors, not necessarily confined to simply an initial violent expansion followed by gradual deceleration, and, in fact, the response to such an initial impulsive input, I should think would be quite complicated. In other words, the current apparent slow-down, followed by an apparent speeding up of universal expansion in this picture is not to be viewed as unexpected or anomalous but rather just part of a complicated spectrum of characteristic behaviors, something of which we have so far seen just a glimpse.

And, if there are indeed other universes, the inner workings of U and the transmissions it returns would not necessarily be the same for all of them. Finally, the way feedback systems work is by correcting the "errors" that appear at that summation sign we see in the above diagrams. Generally there are disturbances that come into systems like that usually including a spectrum of what is often called random noise and system designers have to take those kind of things into account. It seems to me that in our situation spacetime itself would contribute some such behavior, including quantum effects depending on ambient temperature and I will talk about that a bit in Chap. 42. In the meantime, let's see what Chap. 40 has to say.

40

The Alternative Model and Quantization

Perhaps a few words about the relationship of the alternative model to quantum considerations might not be inappropriate at this point. As mentioned in Sec. III, this book is not concerned with kinematics and therefore with the quantum mechanical uncertainty relationships of the associated canonical complementary dynamic variables. Nevertheless, there are some aspects of the model that coincide with the basics of quantum mechanics. To begin with, right up front we can cite the fundamental fact that the twist of a torus knot is quantized in the same sense that the De Broglie–Schrödinger electron's orbital waveform is quantized. Also we recall that, as the outer product of two vectors (whose components are fermions), the boson matrix, M, constitutes an ***operator*** that produces a set of states in the next higher fusion process. As we have seen, in detail (see Sec. II), that process is inherently statistical and its analysis calls for a quantum mechanical-like treatment of the associated degenerative states that can result.

Following up on that, we note that from a fundamental point of view, a salient feature of quantum mechanics is its formulation in terms of ***complex*** algebra and, as we saw in Sec. III, the alternative model lends itself to such a formulation in a unique way, namely in terms of the relationship between twist and charge. The model starts out with a high degree of symmetry, actually the bilateral antisymmetry of its four basic FMS. Together with the symmetry-breaking choice of a traverse direction, upon fusion the result is an ***orthogonal*** relationship between the gradients of charge and twist (the surrogate for isospin), which, in turn, implicates the formulation of a complex state vector (with twist as the real part and charge as the imaginary part) as the phase of an exponential state function. Also, we note again, the correspondence to the canonical ***space and momentum*** variables and their combination in what amounts to a version of the Poisson bracket formulation, another subject for later publication.

Another point of contact, this one more explicitly with the "old" quantum theory of Bohr *et al.*, emerges, we recall, from Einstein's "ansatz", the one that maps the quantized trajectory of a particle in a central field of force onto a torus (see Sec. V). In fact, Einstein reformulated the Bohr–Sommerfeld quantization criterion into one which, in terms of our model, calls for separately quantizing, in multiples of the Planck constant, the integral of momentum along both the meridianal and longitudinal directions of the torus. But in Sec. V, we considered the ***converse***: point of view in which the solitonic deformation of an MS is viewed as a "***particle***" moving along its twisting torus knot locus under the influence of an ***implicit*** central field of force. Then, combining Bohr and Einstein, we explicitly introduced such a field in the form of the Bohr formalism for the hydrogen atom but modified to express a gravitational rather than an electromagnetic effect. That was mainly all directed toward the formulation of our "indigenous parallel" to the Higgs potential.

However, as per our admonition above, we recall that the resulting formalism for our solitonic particles contains ***Planck's*** constant, $\hbar$! And the speed of light, c! And the gravitational "constant", G! All of which suggests we start a new chapter, one explicitly concerned with a possible connection between gravitational and quantum effects. So that is the subject of Chap. 41 below. In the meantime we note that, as experience tells us, at small enough scales, quantum effects begin to emerge as attested to by the discrete line spectra of atomic activity. Thus, postulating that our basic solitonic particles can be viewed as much smaller entities circulating around under the influence of a central field of force implies the existence of ***discrete orbits*** in analogy with what is observed on the ***atomic*** scale and that led Bohr to the "old" quantum theory.

Which suggests that carrying out a "new" quantum analysis of this phenomenology, say in terms of the Schroedinger equation or its equivalent, can answer the questions raised in the previous two sections about the relative orbits and masses of the electron and the nucleons, something that, unfortunately, has not yet been performed. The closest we came was to generate that "hyperbolic" constraint relationship, in Sec. V between the electron and nucleon toroidal radii, something reminiscent of the Heisenberg uncertainty relation.

On the other hand, it cannot be overlooked that, at least as treated in our model, an MS is not an inherently quantum object, except insofar as the appearance of Planck's constant in the formalism makes it so. Otherwise, we have here an ontologically ***classical*** entity that manifests the quantum mechanical concept of ***spin*** in multiples of 1/2, and that features a discrete ***charge*** and a quantized ***twist***, intimately related to the notion of ***isospin***. So, with the expected completion of a "Schroedinger type" analysis we should have a full quantum picture. At that point the alternative model will have removed a long-standing boundary by providing an epistemological ***bridge*** between the physically ***comprehensible*** ontology of those entities fundamental to matter — especially in regard to their general relativistic relationships — and their quantum mechanical analysis, a bridge sometimes characterized as being fundamentally impossible to build (but see Chap. 41, below). Actually, the emergence in Sec. III of a classically ontological ***rationale*** for the use of complex, exponential notation is an example of that "epistemological bridge".

In a larger sense, this point of view follows in the tradition of the Einstein, Bohm and De Broglie "***realistic***" view of fundamental physics as being ontologically comprehensible, as contrasted with the dualistic, irreducibly ***paradoxical*** Bohr, Heisenberg outlook. All something to look forward to, but there is one aspect of that dichotomy that bears mention herein. It has to do with the fields one associates with elementary particles in the SM. As noted above, Quantum Field Theory views particles as quanta, each of its own field. In contrast, Bohm/De Broglie view fields as ***accompanying*** particles. In that regard we note a discussion in [10] (but not reproduced herein) of how such fields might be ***generated*** by fermions, on the one hand, and by bosons on the other and how there may be a connection therein with Pauli exclusion.

41

Quantum Gravity?

In 2005, to commemorate the 100th anniversary of Einstein's "miraculous year", World Scientific published a book entitled "100 Years of Relativity" with the subtitle "Space-time structure: Einstein and beyond", ed. **Abhay Ashtekar** [66]. It will probably come as no surprise that the emphasis on the "beyond" is essentially on the unification of General Relativity and Quantum Mechanics. The book is divided into three parts and a total of 17 chapters each by a different author with seven chapters devoted exclusively to the unification of General Relativity and Quantum Mechanics, often viewed as ***the*** piece that would make the jigsaw puzzle of elementary particle physics complete after its partial unification by Yang–Mills theory and its extensions (plus, of course, the Higgs contribution).

The first part, consisting of a single chapter is entitled "Development of the Concepts of Space, Time and Space-time from Newton to Einstein". Of particular interest to this book is the way the author of the chapter, **John Stachel**, summarizes that development in terms of a diagram he calls "the Bronstein cube" (see Figure 41-1). Arrayed on the corners of the cube are eight theoretical points of view and a way to progress from one to another via three orthogonal directions each of which is associated with a fundamental constant that distinguishes the two. For all such progressions, the starting point is the Galilei–Newtonian Theory. For instance, the kinematics of Special Relativity is distinguished from the old Galilei–Newtonian kinematics by the invariance of the speed of light to the velocity of inertial systems while Quantum Field theory (which is relativistic) involves both Planck's constant and an invariant speed of light. But the ultimate point of view, Quantum Gravity, as the theoretical point furthest from the old Galilei–Newtonian theory, and differing from it by all three constants, c, G and h, can, cubewise, be reached from it by six, three-leg paths (count them) along each of which all three constants appear.

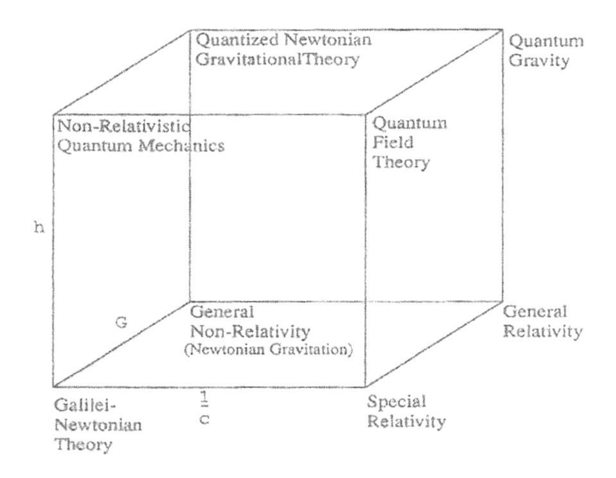

Figure 41-1. The Bronstein cube.

Stachel then selects the two approaches to quantum gravity that are currently considered (or at least most mentioned as) the most likely to get there, Loop Quantum Gravity and String Theory and shows a particular path for each, namely the path that best illustrates that approach. String theory was discussed in Sec. VI and is also the more widely known and discussed of the two.

To be fair, we ought to give Loop Quantum Gravity some space here. It turns out that Prof. Ashtekar, who, as noted above, is the editor of the cited book and also one of the founders of LQG theory and a principal contributor to its formal basis, wrote a chapter in the book entitled "Quantum Geometry and its Ramifications" in which he expounds at some length about it. I can do no better than to relate the professor's exposition thereto so here, (considerably compressed) is what he has to say in terms of an introduction to what LQG is about; First:

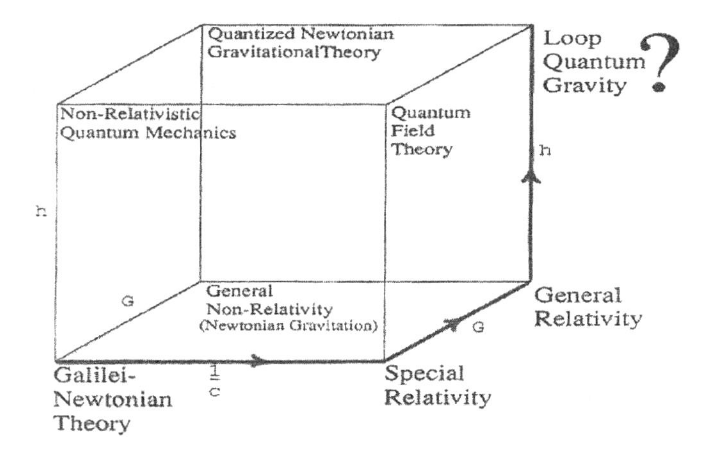

Figure 41-2. The LQG path.

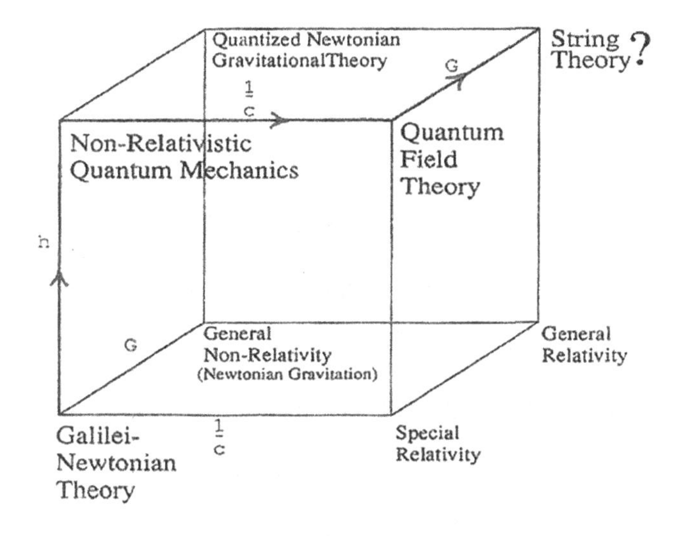

Figure 41-3. The string theory path.

"The central lesson of general relativity is encoded in space-time geometry."

He goes on to note that there are spectacular predictions associated with GR but also limitations such that

"Matter fields become singular but so does geometry."

Correspondingly, therefore:

"The key idea at the heart of loop quantum geometry is to retain the interplay between geometry and gravity but overcome the limitations of general relativity by replacing classical Riemannian geometry by its suitable quantum analog — There is no background metric, no passive area on which quantum dynamics of matter is to unfold — in striking contrast to approaches developed by particle physicists where one typically begins with quantum matter on a classical background geometry and uses perturbation theory to incorporate quantum effects if gravity."

Then, in more detail: "The basic gravitational configuration is an SU(2) connection, A_a^i, on a 3-manifold, M, representing 'space'. As in gauge theories, the momenta are the 'electric fields', E_i^{acd}. However, in the present gravitational context, they also acquire a space-time meaning: they can be naturally interpreted as orthonormal triads (with density weight 1) and determine the dynamical Riemannian geometry of M" — "The basic kinematic objects are:

(i) holonomies $h_e(A)$ of A_a^i which dictate how spinors are parallel transported along curves or edges e and
(ii) fluxes $E_{s,t} = f_s t_i \, E_i^a d^2 S_a$ of electric fields E_i^a (smeared with test fields t_i) across a 2-surface. The holonomies — the raison d'etre of connections — serve as the 'elementary' configuration variables which are to have unambiguous quantum analogs" — "The first step in quantization is to use the Poisson algebra between those configuration and momentum functions to construct an abstract (star-) algebra A of elementary quantum operators. This step is straightforward."

Did you get that? I know, I know; it is more than a little arcane! (and my presentation of it probably did not help much either. So I must appologize to the reader as well as to the professor Ashtekar). Nevertheless, I have some comments about what you just read. The first comment is that, if Professor Ashtekar says something is straightforward, it is straightforward. The second comment is that, as per my emphasis throughout this book, our

Alternative Model shares the fundamental LQG point of view of eschewing a background metric; we also have "no passive area on which quantum dynamics of matter is to unfold". And, as you have probably gathered, we also feature an SU(2) connection as does LQG. Also, although it is not immediately obvious from the above, our "configuration and momentum" variables, the twists and charges of our particles, are quite analogous to those of LQG. Furthermore, as per Sec. III, in effect we "use the Poisson algebra between those configuration and momentum functions" to construct an algebra, although admittedly, not the kind of algebra Ashtekar has in mind.

On the other hand, while our Alternative Model is all about a set of basic particles and its ramification, LQG is completely preoccupied with what Ashtekar characterizes as "replacing classical Riemannian geometry by its suitable quantum analog". In other words, modifying the description of the space that General Relativity talks about. As a result, just like String Theory, LQG is concerned with what might take place around the Planck scale of distance. However, the actual treatment is fundamentally different; in LQG there is much work devoted to spin phenomenology but it is associated not with the spin of particles as with the Standard Model or even our Alternative Model version but of links in a network of which space is postulated to be composed.

And here, I believe, is somewhere that LQG and our Alternative Model can find common ground. I refer to the dynamic behavior at the smallest scales of the spacetime from whence our AM particles arise as solitonic disturbances, ***perhaps*** treated as a spectrum of random interference, including quantum effects, for the torus knot generators as mentioned in the previous chapter.

In any event, by now an immense amount of work has been and is being devoted to LQG and a rather serious competition exists between the LQG and String Theory camps in their quest for ultimate enlightenment. In a way, both encampments have also taken what might be characterized as the reductionist route, not by postulating a further level of particulate matter as in the SM, but rather by a reduction of the scale with which they are concerned.

Of course, that reduction has been balanced by a corresponding expansion of the amount of brainpower devoted to the incredible mathematical intricacies encountered along the way to said ultimate enlightenment. To say nothing of the associated expenditure of monetary resources, so I will not, although not everyone has been so circumspect [67].

Which looked like a good place to end this chapter, but I had forgotten something; we need to go back for a minute to those Bronstein cubes and the c, G and $\hbar$ traversals that span them. String Theory and LQG are not the only players on that pitch; our Alternative Model has something to say about the game we play thereon. To begin with, you may recall how we endowed our Möbius strips with mass in Sec. V: we modeled them as concatenations of Torus knots whose mass, our physics told us, involved the local stress energy density, A which, translated into a gravitational "constant", brought in both G and c.

Then, in a later chapter, having to do with the construction of a "parallel" to the Higgs field, we played around with the "old" Bohr atomic theory, as applied to a long-forgotten, but most imaginative Einstein notion, with the result that naturally introduced the $\hbar$ into

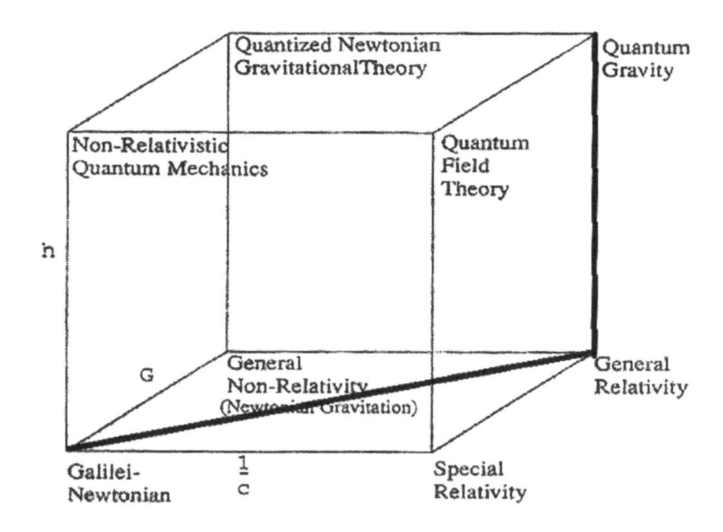

Figure 41-4. Cube traversal for the Alternative Model.

the picture. In other words, we traversed the BC along a path less — in fact, probably never before traveled as in Figure 41-4. Nevertheless, it starts and ends where the other two do! Now, some purists may take issue with me for playing fast and loose with the gravitational "constant" in the interior and on the periphery of my particles. To which my sanitized response would amount to something like "well, they are my particles and I can treat them to gravity any way I like! And besides my particles ***have*** an interior and a periphery and yours do not. So there!"

But, seriously, as I noted back in Sec. V, nobody really knows how gravity behaves down at those miniscule scales. Maybe continued intensive probing of nucleon interiors will shed some light thereto. We shall see, or may be not. Your chances are probably better than mine.

42

Conclusions

Although we had a chapter (Chap. 34 in Sec. VI) on our model's relationship to String Theory, we probably did not spend as much time on it as we should have so, to make up for it, I would like to begin these concluding remarks with this profound assessment excerpted from a 1989 paper by Edward Witten [68]: "In my opinion, the basic challenge in string theory is not, as sometimes said, to 'understand nonperturbative processes in string theory'. In fact, the basic problem does not lie in the quantum domain at all. The basic problem, now and for many years to come, is to understand the classical theory properly." And further: "Understanding the classical theory means above all understanding the geometrical ideas that parallel those of general relativity —"

Great! In my book (as they say) Witten is not only a phenomenal physicist and mathematician but a straight thinker (!). So it is my fond hope and expectation that this book addresses his challenge, not only as explicitly stated, but as it might concern a larger context and in ways that, as implied in the Introduction, do, indeed, offer a more efficacious, more systematic approach to the development of a model of the elementary particles of physics. I would welcome any and all comments in that regard.

At the same time, there has always been a question lurking over the model's development: "is it real?" Quite a long time ago, a well-known expert on Quantum Field Theory, String Theory and particle physics who chaired a symposium session on "Further out" ideas was kind enough to review a copy of the summary of my model as it existed at that time. His conclusion was that the model appeared to replicate the Standard Model's algebra (which was of course what I was trying to convey) but was nevertheless irrelevant! Of course, the encounter took place well before much of the detailed taxonomical discussion included herein was available or what is, arguably, especially relevant to the question of reality, well before the work on the solitonic nature of the model's "particles" had been carried out to say nothing of the various topics that emerged therefrom. What the reviewer also missed was that the AM replicates the SM algebra in a geometrical, visualizable, notably efficient way that manifests the self-containment epistemology of Sakata and the Fermi–Yang notion of bound states. I like to think that at this point in time he would come to a much better assessment.

But, before we go too far along in such comparisons, let us not forget the admonition of the Preface: we are dealing in allegory here and all our models are just that, whether SM, AM, or any kind of M. That being the case, a necessary requirement for evaluating a model is self-consistency. Equally important, however, is the minimization of external assumptions. The famous mathematician John von Neumann is reputed to have claimed (presumably at least partially in jest) that with some minimal number of free parameters he could model an elephant! And with one or two more he could make its trunk wiggle!

I would not go quite that far but, all things considered, my current assessment is that the Alternative Model described herein is as "real" as the Standard Model whose taxonomy, attributes and interactions it claims to replicate but minus the latter's quarks, gluons, "color" or the *ad hoc* importation of a Higgs field. Or the importation of from 17 to, possibly, a couple dozen experimentally determined values of the parameters necessary to complete the model's description. And I would hope that this last remark is not construed as a criticism of the SM which, as I have stipulated in the preceding, is a truly remarkable edifice.

Nevertheless, the remark is factual and it has implications. To explore them a bit perhaps I might dredge up a certain folk maxim that has attained currency of late as a way to cope with the exigencies of life, namely "***It is what it is***". You may have heard it — I have used it myself quite a bit. The way I understand it, what it means is that when faced with a situation with which one has no hope of coping, the only way left is just to accept that fact, to incorporate its reality, and then to just go on with one's affairs. In that regard, the custodians of the Standard Model, have, it would appear, simply accepted the fact that it has no hope of explaining the raison d'etre of those 17 or 20 or whatever parameter values — they are what they are (!) — and have, indeed simply accepted them as given and gone on from there to look beyond the model for answers. But now, the Standard Cosmological Model, needing to include a model of the elementary particles in its set of basic "givens" has, naturally, just incorporated the SM's model! Would you not? In other words, the cosmologists assume that the model "is what it is" ***because*** it has already been deemed to "be what it is".

Summing up, we may have realized here the elevation of a maxim from its residence in the folk arena into the realm of sophisticated logic, and it now reads "it is what it is ***because*** it is what it is". Or, in initials $I^2WI^2BI^2WI^2$. Kind of exciting, don't you think? Let me know.

But, enough of this tomfoolery; I was only kidding. But, was I? Well, let's go back for a moment and take another look at our old friend Ouratio the Ouraborus and how he is doing after that rather extreme procedure. Now, on the one hand, it left him without a head, a most awkward situation. On the other hand, because we had retraced the expansionary history of the Universe, it had left him on a par with, and able to mate with his tail rather than trying to swallow it! We ended up with a loop, that is a re-entrant process and, it seems to me, a manifestation of $I^2WI^2BI^2WI^2$, or to put it another way, an eigenform, in this case, the set of elementary particles as per Kauffman's box procedure and closely related to my long-ago notion of the connection between the universe and its constituents.

Of course, the loop is not the Universe. It's not even a representation of the Universe. But it is kind of representative of the ***story*** of the Universe. The two rise and fall together. And personally, I don't believe either will ever disappear. Or, that there has to be a cataclysmic beginning or an ever-lasting expansion. However, I do believe that expansions and contractions are inevitable, in fact a whole spectrum of them, something I talked about before as you may recall. But an inflationary expansion of some kind is not to be ruled out. In fact, it's a necessary ingredient of my story (in Chapter 3) of the loop's creation.

Finally, speaking of spectra, you may also recall the small scale part of the universal spectrum I mentioned briefly in Chapter 3, wherein we see all kinds of topological configurations of space including the little "hill and valley" characteristic required for particle creation, our Alternative Model's parallel to the Higgs potential. You may also be familiar with the way cosmologists talk about how universal expansion begins [1]. What they say is that the universe starts out poised in a precarious, only conditionally-stable position atop what they call "the false vacuum" of spatial potential energy and sooner or later it begins to roll or fall down toward the "real vacuum" picking up steam in what amounts to the inflationary phase. With apologies to the cosmological community on the one hand and the particle community on the other, I can't see the difference between the "false and real" vacua associated with Universe creation, the Higgs potential associated with Standard model particle mass and the "Hill and Valley" spatial characteristic of Alternative Model particle mass. The way I see it, they're all manifestations of that little "theorem" I talked about in Section V, the one that was concerned with particle differential geometry. Remember that?

So; where are we? It seems to me that, to some extent, all's reasonably well with the world; our particles must emerge as expected as long as some kind of inflationary scenario is included as part of the SCM and the custodians of the SM can relax, also to some extent, because the set (but not the nature) of the particles they find "are what they are". They still have a bunch of parameter values they don't know how to justify on a more fundamental basis. And, according to our Alternative Model, they're still working too hard and need to forget about Quarks, Gluons, Chromodynamics, SU(3) color and all those fields, at least as currently conceived. And they ought to take a long look at particle geometry. But Yang–Mills/Gauge, and Noether theories are fine and so are "families" of particles as discussed in this book!

You know, on the whole the universe is a cold and inhospitable environment punctuated by contracted regions of mind-bogglingly intense activity describable mainly in statistical terms. Nevertheless it harbors an incredible array of diverse organization, all seemingly flouting the edicts of thermodynamics, "a hierarchy of structure" to quote Vilenkin once again, from our life forms here on earth to the farthest reaches of the cosmos. And ultimately all dependent on the "elementarity" which, as per the thesis of this book, is provided by the micro, micro deformations that exist in and of the localized spacetme of our elementary particles. Perhaps you recall John Wheeler's recollection of Albert Einstein's "vision of a totally geometric world, a world in which everything was composed ultimately only of spacetime". It seems to me that *our* ***Alternative Model*** *of* the elementary particles constitutes *a basis for the* ***manifestation*** *of that vision.*

43

Supplementary Conclusions

What has also emerged but very late in the development of this book appears to be a *universally applicable* ***principle*** which in application may come in more than one form and will take a while to explain. If you recall, we began this journey by talking about (2, n) ***torus knots*** and ***two-strand*** braids with closure, both of which may be viewed as forming a set of ***four*** rudimentary ***Möbius strips***, three of which (according to the thesis of this book) constitute the basis for a taxonomy and, ultimately, of all the atoms in the periodic table, those atoms that underlie the consequent physics, geology and chemistry of Planet Earth and even the biochemistry of life itself, based, as it is on the geometry of those ***two-strand*** braids of DNA (Recalling Flapan's "Moebius Ladder") and the ***four*** kinds of molecules that connect them! And another, overlapping, ***complementary*** set of three MS that (according to the speculative discussion in Section VI) constitutes a complete, ***complementary*** universe identified (in that section) as Dark Matter.

Talking about elementary particles and the biochemistry of life in the same breath highlights what may seem to be suspiciously like "coincidences". Or is there more to it than that? Good question. However it was only after what I had originally considered to be a complete manuscript and had submitted it to the editors to be turned into presentable book form that I finally realized I had been remiss in not following up on that question. That is, on the fact that (1) ***both*** our Alternative Model of the elementary particles and DNA, the genetic blueprint for the life forms of much of the planet involve ***four*** elementary entities, (2) both involve ***two-strand*** structures and (3) both involve the ***breaking of mirror symmetry***, something that, in the particle model, carries through all the way to the violation of ***CP conservation*** in Beta decay and more, including the two universe hypothesis. Well, so as not to upset the orderly (and timely) development of the manuscript into a book, I decided to postpone the following-up process to an appendix (Appendix F. "The alternative model, DNA, and Dark Matter Redux").

Nevertheless, it's still important and, as it turns out, the results are most edifying. First there is the cited Principle; it is simple and straightforward in its bare concept but embodies augmentation in many applications (see below). But with or without augmentation it appears to underlie the organization of structures ranging in scale from that of the elementary particles, to the Double Helix of DNA, to the universal population of the cosmos by Vilenkin's "hierarchy of structure", and naturally, evokes questions as to why that might be so. So, to begin the explanation referred to above, suppose we take a look at what emerged from Appendix F. In summary, ***both*** DNA and our Alternative System of elementary particles consist of ***four*** basic entities. each describable by a ***three***-letter word with each letter taken from a ***binary*** alphabet. Furthermore, the four entity basis in each case is ***separable*** into two ***complementary*** bases. All of which leads to an ***algebra*** that describes both cases! Most gratifying; it's not every day that one witnesses a correspondence between the description of organizations so diverse in scale as elementary particles and DNA!

Like the race to reach the North Pole (or was it the South Pole?) arriving at the true structure of the DNA molecule had been quite a competitive affair. When he and **Francis Crick** won the race, **James Watson** recalls, he felt that even their main competitor, **Linus Pauling**, would have to acknowledge "the overwhelming merit of a self-complementary DNA molecule". One wonders how he would react to being credited with the discovery of a most important biological manifestation of what might constitute a "Universally Applicable" Principle, one that, although it's not the "Theory of Everything", should provide additional guidance in what to look for in our quest for better understanding of our place in the cosmos!

As far as "universality" is concerned we note that, in itself, complementarity can be either ***physical***, i.e. tactile, visible, etc. or ***mental*** — an idea, concept, principle, etc. Another way to say it, somewhat broader but perhaps less definitively separable, is in terms of ***Ontological*** versus ***Epistemological*** points of view, respectively. But, in any event, ***Complementarity*** is ubiquitous and unavoidable in nature, for example, in the body plan we, along with our "higher-creature" relatives, inherit from our "lower-creature" ancestors, including, for example, four limbs, two in "front", two in back, with left and right mirror symmetry and, also, in one form or another, in the dual-gender requirement for reproduction of a species. And two complementary halves to our brains, two eyes, two complementary sets of teeth, upper and lower, two lungs, two chambers to our hearts, etc. Also, our existence on Earth may be conditioned on the complementarity afforded by the rotation of the Earth giving us night and day, and its inclination and trek about the Sun giving us the seasons. It's been said that even the cycle of the moon may have an effect on terrestrial evolution.[1]

Of course, there's more to the ***ontological*** aspects of complementarity than these examples. As we've seen, in DNA and in our Alternative Model for the elementary particles, the two cases compared in Appendix F, complementarity also requires the ***breaking of mirror symmetry***. And in Sec. V we saw how the variation of stress associated with the complementary ***hill and valley*** topography internal to our Alternative Model particles produces the toroidal topology leading to particle formation and persistence and with the choice of a direction of a traverse constitutes a break in symmetry.

Augmentation also comes into play in the examples of Appendix F namely in terms of that ***four, three, two*** plan. Or, in the case of other systems, something akin to it; I haven't really had the

[1]I put that in there just to see if you were paying attention. So go ahead, scoff; can you prove otherwise?

chance to examine other systems in detail for purposes of this book but ***Maxwell's equations*** come to mind as a natural candidate. Right away we see four equations and, on the L.H.S. of the equations, four "entities" — the two field vectors (Electric and Magnetic) and the two mathematical entities that operate upon them (Div and Curl) and, as we know a break in symmetry in terms of the way the electrical and magnetic intensities are influenced by the R.H.S. of the equations. We should note that, as is of course well known, the equations have served as the basis for the theory of electromagnetism wherein the electric and magnetic fields are sometimes expressed as "***dual***" rather than "complementary" entities. The equivalence of the terminology also applies to the upgrading of ***String Theory*** wherein it's separate, ten-dimensional versions are found to be combinable into complementary, "dual" pairs as part of what is now known as eleven-dimensional **M Theory**.

Without going into detail, the generalization of Maxwell's equations in terms of ***Yang–Mills theory*** is another example of complementarity. And consider the ***Dirac theory*** we outlined in Sec. II; it's expressed in terms of a pair of two-component, vector equations operated on by four component, complex matrices and we end up with a particle and a complementary antiparticle differentiated by choice of traverse direction. Also, if you think about it, the concept of ***isospin***, (whether "strong or "weak") is a manifestation of Complementarity, as graphically illustrated in Sec. IV of this book. And, finally, going back to the definitions of our particle model in Sec. II and the speculations of Sec. VI, the splitting of our initial four-component set of basic particles into the bases for two distinct but complementary universes, ours and the "dark one". Actually, in retrospect, the book is awash with complementarity; you can't avoid it because it's built into the basis of our Alternative Model as we saw in Appendix F.

So much for a quick review of just a few of the manifestations of the ontological aspects of complementarity. How about the epistemological aspects? Well, staring us in the face is that existential confrontation we talked about in Chap. 36 between the

Anthropic and Multiverse points of view,[2] one assuming the unique predestination of our species and the other simply an evolutionary process given monumentally large statistical samples over space and time and, by implication, assigns no special importance to our species at all! There are, of course, other ways to frame the situation but it seems to me, at least in terms of the importance to our species, that those two constitute a most succinct summary of the either-or, this-or-that, Yin–Yang nature of, arguably, the most pertinent question facing mankind today in our quest for fundamental understanding of "the meaning of it all": Where do ***we*** fit in the grand scheme of things? And the way I see it, the two points of view are also in a complementary relationship just as much as the pair of universes (ours and the "dark" one) we talked about in Section VI.

On a somewhat less philosophical level, consider Heisenberg's Uncertainty Principle; it puts each of several equivalent pairs of basic complementary quantities into a simple, hyperbolic constraint in terms of the accuracy with which both members of the pair can be measured simultaneously. It follows from the basic formulation of Quantum Mechanics, especially ***Wigner's*** approach [70], which, believe it or not, is found in a different guise in terms of the so-called ***Ambiguity Function*** of radar [71] that pairs range (signal time delay) and Doppler velocity (signal frequency shift) information (or aperture size and beamwidth) in a similar constraint. Actually, the formulations of all such pairings may be viewed as just a natural consequence of the basic tenets of Fourier analysis as it applies, more broadly in the theory of communication (information).

The Dirac theory might also be considered in epistemological terms since it shows that each basic fermion must have an antifermionic partner, in fact a ***complementary*** partner as mentioned above. Which reminds us of how the bosons of Sec. II are also configured as just such a fusion of a fermion and an antifermion. And how, although each member of the pair requires a directional figure to execute two traversals in order to return to a starting point

[2]There are very few loose ends in this book!

with its original orientation, such a figure can get by with only one traversal of a boson. The benefits of complementarity!

And finally, consider the most basic generalization of all, ***Noether's Theorem***, which pairs the ***invariance*** of an entity governed by a symmetry principle to changes of a reference system with the ***conservation*** of an associated dynamic entity within a given system and which according to [72] "forms an ***organizing principle*** for all of Physics". Indeed it does, but I must bring to your attention that the fundamental pairing of invariance and conservation also constitutes an epistemological ***manifestation*** of complementarity! Each member of the pair implies the other: if we find an entity that is invariant to changes in reference, we can predict the existence of an associated entity whose value never changes as measured within a given reference and, conversely, awareness of an entity whose value never changes within a given reference system implies that there exists an associated entity that is invariant to changes of reference. From that point of view, Noether's theorem, the noblest of them all, can itself be viewed as being encompassed within the Principle of Complementarity.

In summary, we can say, I believe, that "In terms of Epistemology, the ***essence*** of ***Elementarity is Complementarity***". If something is elementary — truly elementary — then it is part of a ***Complementary pair***. This is not to say that it may not be augmented, as mentioned above, by other essential elements, perhaps the ***four, three, two*** plan as per the two systems we treated in Appendix F — we need to expand the investigation. But complementarity is ***essential***. Although the precise way in which complementarity is manifested may vary, it may be said to constitute an essential ***Homology***. We saw a number of examples above but there are any number of others begging to be delved into. And I almost forgot one of the simplest and most important for which I offer my apologies to Isaac Newton: Even to this day, Sir Isaac would be pleased to know, ***Action*** still begets ***Reaction***! Which it does in many contexts, even unto basic human relationships and most often accompanied by consequences, including some of the most disastrous one can imagine. Will we never learn!

And, finally, as we all know, a picture is worth a thousand words, so here we have one that may be worth all of the above; I refer to the ancient and honorable icon of Chinese philosophy, the Yin Yang symbol. If anything ***epitomizes*** the notion of complementarity, that's it! So here in basic, simple form it is, elegant in its simplicity.

The basic philosophical notion associated with the symbol is that Yin and Yang can be thought of, not as being in opposition but as ***complementary*** forces that interact to form a dynamic system for which the whole is greater than the parts and, further, that ***everything*** has both yin and yang aspects. Here I would like to call your attention to a more mundane observation: that the curved boundary is reminiscent (at least to me) of our "hill and valley" characteristic of that little "not insignificant" theorem of Section V, the one that underlies both the Higgs potential of the Standard Model and our Alternative Model's "parallel" to it.

We could go on citing the widespread wonders of Yin–Yang but recently a related subject caught my attention. It's taken from the world of recreational geometry and is known as Penrose tiling (because it was invented by the world-famous British mathematician and Physicist **Roger Penrose**); you may already know all about it. My reference here is a recent Scientific American article [73] although I've seen other references to it through the years. I like to think about its' realization as a kind of basic Yin–Yang diagram but using straight lines instead of a circle divided by that doubly-curved partition. In the figure below we see a canted parallelogram, with (necessarily) two equal acute and two equal obtuse angles, divided by two shorter lines whose included angle is the same obtuse angle. The result is to destroy the mirror symmetry of the original parallelogram and divide it into two ***complementary*** figures that in the reference article are known as a "kite" and a "dart" for obvious reasons.

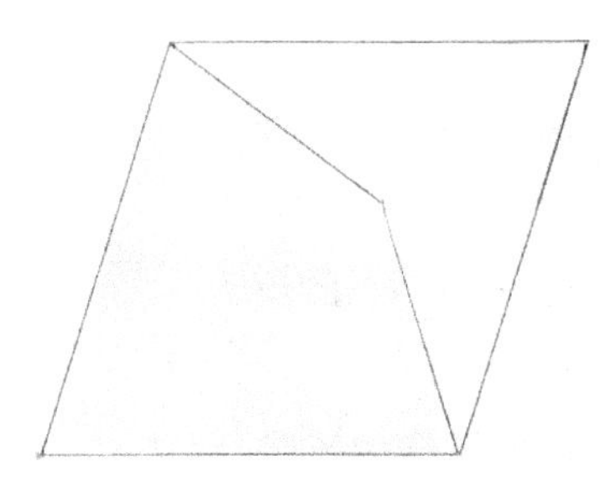

The article cites the way Penrose found how to combine those two (or similar pairs of) figures in various ways so as to completely tile a flat surface without gaps and no particular pattern repetition. It also shows a particular example (not shown here) of such a tiling from which we can deduce particular values for the angles, namely 72° for the acute angles and 108° for the obtuse angles. It goes on to cite a number of extensions of Penrose tiling for example to the structure of what are known as "quasicrystals" and to the phenomenon of self-symmetry — the organizational similarity observable at different scales of some phenomenon (as in fractals). The point here is that ***complementarity*** is really a very simple concept that underlies a world of phenomena and ideas.

Well, one thing leads to another but for now we've probably carried complementarity as far as we need or ought to in this book. It may be that another book is called for just to explore the ramifications of our principle and its augmentations (including those suggested above). And with that ambitious look ahead, it looks like ***this*** book, except for the Appendices, (don't forget Appendix F!), the References and the Index, is coming to an end. All things considered it's not a bad book; one of the main things I like about it is that its progenitors were so correct, so prescient; Clifford was eminently so; as were Kelvin, Sakata, Fermi and Yang, Einstein, and Bohr. Kauffman (Chap. 39) is as well and so is Witten (above) and it's been my privilege to put something together that recalls, and in some way manifests and validates the collective vision of those visionaries.

So in closing, I would like to say that it's been a pleasure talking to you and having had the opportunity to describe what's taken up much of my time at a phase in life where many others in similar situations would have spent the time in more important things like travel, fishing, golf, bowling, folk dancing, etc. — I'm sure you can fill out the list but the truth is that I have no regrets on that score! I've been quite busy, have learned a lot and kept out of trouble However, I do hope my modest contribution, my "footprint" to quote Eddington (remember that quotation at the beginning of the Preface) will make some difference in the way people think about particle physics and, perhaps of such endeavors altogether. I leave you with another Eddingtonian pronouncement for your continued cultural edification:

> *"It is one thing for the human mind to extract from the phenomena of nature the laws which it has itself put into them; it may be a far harder thing to extract laws over which it has no control. It is even possible that laws which have not their origin in the mind may be irrational and we can never succeed in formulating them."*
>
> (Eddington, "Space Time and Gravitation")

> "*Something else to think about but don't worry about it, just yet, anyway*"
> (Jack Avrin, 2014)

44

Postscript

Well, the book finished with a bit of a flourish, uniting the seemingly disparate formalisms of the Alternative Particle Model with DNA, that well known, two-strand format upon which we depend for the perpetuation of our species and making possible the emergence of a unifying principle. Of course that two-strand format can actually result, they say, in a potentially astronomical array of life forms whereupon the process of evolution culls that down to only those that can survive in the support system provided by the environment. But a strange, paradoxical thing has happened on our planet: the evolutionary process has produced a certain cerebrally-advanced life form (that's one way to describe us!) that can actually modify its environment and in ways more conducive to its own and *only* its own survival and on such a grand scale that another law of nature seems to have taken over; it's called the law of unintended consequences and, sure enough, things have turned out badly, very badly indeed in that environment, which, as we all know but don't all want to admit, has begun to fight back. Of course, the immense problems, indeed potentially catastrophic problems that emerge lie outside the domain of this book but they have been stressed over and over by immensely qualified people. And the message that comes through, over and over is just this: what we desperately need, and quickly, are radical changes in the way we act, the way we live, and what we view as important in life, as the dominant species here on this tiny blue planet in the immensity of space and time. What worries me is that, as a species, we may not be so constituted as to realize the unanimity of purpose that could make the necessary changes possible. And that all the grand pursuit of knowledge, our quest for the enlightenment and understanding I talked about early on may be all for naught, an outcome too heartrending to contemplate. Let us all clasp hands and pray that it be not so. And perhaps raise our voices to help fend it off. Or even do some fending!

Appendices

Appendix A

Illustrating The Double Traversal Requirement

TRAVERSE AROUND NHT = 3 FMS

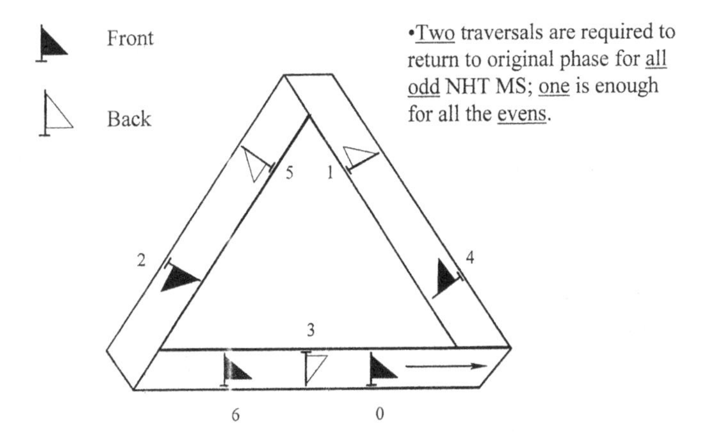

Appendix B

Torus Knot Connection Terms

Beginning with the formulas for the CS of the first kind, Eq. (21-13)

$$\Gamma_{abc} \equiv \left(\frac{\partial^2 \mathbf{S}}{\partial u^a \partial^b}\right)\left(\frac{\partial \mathbf{S}}{\partial u^c}\right), \quad a, b, c = 1, 2 \tag{B-1}$$

and for the torus knot location vector, Eq. (21-1)

$$\mathbf{S} = \hat{\imath} w \cos\phi + \hat{\jmath} w \sin\phi + \hat{k} r \sin\theta, \tag{B-2}$$

the indicated derivatives are

$$\begin{aligned}
\frac{\partial \mathbf{S}}{\partial \phi} &= -\hat{\imath}(R + r\cos\theta)\sin\phi + \hat{\jmath}(R + r\cos\theta)\cos\phi, \\
\frac{\partial \mathbf{S}}{\partial \theta} &= -\hat{\imath} r \sin\theta\cos\phi - \hat{\jmath} r \sin\theta \sin\phi + \hat{k} r \cos\theta, \\
\frac{\partial^2 \mathbf{S}}{\partial \phi^2} &= -\hat{\imath}(R + r\cos\theta)\cos\phi - \hat{\jmath}(R + r\cos\theta)\sin\phi, \\
\frac{\partial^2 \mathbf{S}}{\partial \theta^2} &= \hat{\imath} r^2 \cos\theta\cos\phi - \hat{\jmath} r^2 \cos\theta \sin\phi - \hat{k} r^2 \sin\phi, \\
\frac{\partial^2 \mathbf{S}}{\partial \phi \partial \theta} &= \hat{\imath} r \sin\theta \sin\phi - \hat{\jmath} r \sin\theta\cos\phi = \frac{\partial^2 \mathbf{S}}{\partial \theta \partial \phi}.
\end{aligned} \tag{B-3}$$

Consequently we find that

$$\begin{aligned}
\Gamma_{\phi\theta\phi} &= \Gamma_{\theta\phi\phi} = -r(R + r\cos\theta)\sin\theta = -rw\sin\theta,\\
\Gamma_{\phi\phi\theta} &= r(R + r\cos\theta)\sin\theta = rw\sin\theta, \qquad \text{(B-4)}\\
\Gamma_{\phi\theta\theta} &= \Gamma_{\theta\phi\theta} = \Gamma_{\phi\phi\phi} = \Gamma_{\theta\theta\phi} = \Gamma_{\theta\theta\theta} = 0.
\end{aligned}$$

Also, using

$$g_{ab} = \left(\frac{\partial \mathbf{S}}{\partial u_a}\right) \cdot \left(\frac{\partial \mathbf{S}}{\partial u_b}\right) \qquad \text{(B-5)}$$

for the metric components we find

$$\begin{aligned}
g_{\phi\phi} &= (R + r\cos\theta)^2 = w^2,\\
g_{\theta\theta} &= r^2, \qquad \text{(B-6)}\\
g_{\phi\theta} &= g_{\theta\phi} = 0,
\end{aligned}$$

whereupon

$$g = |g_{ab}| = (R + r\cos\theta)^2 \mu^2 r^2 = (\mu r w)^2. \qquad \text{(B-7)}$$

Then, using

$$\begin{aligned}
g^{\phi\phi} &= \frac{g_{\theta\theta}}{g},\\
g^{\theta\theta} &= \frac{g_{\phi\phi}}{g}
\end{aligned} \qquad \text{(B-8)}$$

and

$$\Gamma^d_{ab} \equiv \Gamma_{abc}\, g^{cd}, \quad a, b, c, d = \phi, \theta \qquad \text{(B-9)}$$

we find the CS of the second kind to be

$$\begin{aligned}
\Gamma^\theta_{\phi\phi} &= \left(\frac{w}{r}\right)\sin\theta,\\
\Gamma^\phi_{\phi\theta} &= -\left(\frac{r}{w}\right)\sin\theta.
\end{aligned} \qquad \text{(B-10)}$$

Appendix C

Curvature for the Lagrangian

We begin with the RCT (see Eq. (23-11)) of Sec. V

$$R_{\mu\kappa} = R^{\lambda}_{\mu\lambda\kappa} = \left(\frac{\partial \Gamma^{\lambda}_{\mu k}}{\partial x^{\lambda}} - \frac{\partial \Gamma^{\lambda}_{\mu\lambda}}{\partial x^{\kappa}}\right) + \left(\Gamma^{\eta}_{\mu\lambda}\Gamma^{\lambda}_{\kappa\eta} - \Gamma^{\eta}_{\mu\kappa}\Gamma^{\lambda}_{\lambda\eta}\right). \quad \text{(C-1)}$$

The second parenthesis turns out to contribute nothing as a result of its symmetries and the only non-zero terms remaining are for $\mu = \kappa = \phi, \lambda = \theta$ which produces

$$\begin{aligned} R_{\phi\phi} &= -\frac{\partial \Gamma^{\theta}_{\phi\phi}}{\partial \theta} \\ &= \left(\frac{w}{r}\right)\cos\theta. \end{aligned} \quad \text{(C-2)}$$

and for $\mu = \kappa = \theta, \lambda = \phi$ which produces

$$\begin{aligned} R_{\theta\theta} &= \frac{\partial \Gamma^{\phi}_{\theta\phi}}{\partial \theta} \\ &= \left(\frac{r}{w}\right)\cos\theta. \end{aligned} \quad \text{(C-3)}$$

What we want is Eq. (43-10) for the Curvature Scalar, namely

$$R = g^{\phi\phi}R_{\phi\phi} + g^{\theta\theta}R_{\theta\theta}, \quad \text{(C-4)}$$

so that, using

$$g^{\phi\phi} = \frac{1}{w^2},$$
$$g^{\theta\theta} = \frac{1}{r^2},$$

as per the preceding, Eqs. (C-3) and (C-4) sum to

$$R = \left(\frac{2}{rw}\right)\cos\theta. \tag{C-5}$$

Appendix D

Contingency in Second Order Fusion

As mentioned in Sec. II, the salient feature here is that the junctions available for the second order fusion depend on junction selection in the first fusion. The results of analysis of that selection process are repeated here for reference as in Table D-1, which lists half of the second order permutations with coefficients that embody contingencies and Figure D-1, (half of) the resulting degeneracy table.

The situation here is similar to the case of first order fusion except that the coefficients associated with the constituents of the input columns are the ***available*** antiquirks — that is, those left over from the first order fusion process, because they did ***not*** form junctions. With that in mind, we have reproduced below the list of first order fusion words with the junctions that can be formed and added a column with the corresponding available antiquirks. The genesis of the added column will also be formalized but first we discuss it informally. For example, in the case of BB^*, if an x junction is formed, the B^* FMS has both a d^* and a u^* antiquirk left over and since four x junctions can form, we have the term $4(d^* + u^*)$. On the other hand, if a y junction forms, two d^* antiquirks are left over so we have the term $2d^*$. Summing we have the term $4(d^*+u^*) +2d^*$ as shown in the figure.

Table D-1

$AA^*A\ (x)(2x^*)$	$BA^*A\ (2x)(2x^*)$
$AA^*B\ (x)(4x^*)$	$BA^*B\ (2x)(4x^*)$
$AA^*C\ (x)(2x^*)$	$BA^*C\ (2x)(2x^*)$
	$[AA^*D]$
$AB^*A\ (2x)(x^*)$	$BB^*A\ \{(4x)(x^*) + (y)(4x^*\}$
$AB^*B\ (2x)(2x^* + y^*)$	$BB^*B\ \{(4x)(2x^* + y^*) + (y)(4x^*\}$
$AB^*C\ (2x)(x^* + 2y^*)$	$BB^*C\{(4x)(x^* + 2y^*) + (y)(2y^*)\}$
$AB^*D\ (2x)(y^*)$	$BB^*D\ (4x)(y^*)$
$[AC^*A]$	$BC^*A\ (2y)(x^*)$
$AC^*B\ (x)(2y^*)$	$BC^*B\ \{(2x)(2y^*) + (2y)(2x^* + y^*)\}$
$AC^*C\ (x)(4y^*)$	$BC^*C\ \{(2x)(x^* + 2y^*) + (2y)(x^* + 2y^*)\}$
$AC^*D\ (x)(2y^*)$	$BC^*D\ \{(2x)(2y^*) + (2y)(y^*)\}$
$[AD^*A]$	$[BD^*A]$
$[AD^*B]$	$BD^*B\ (y)(2y^*)$
$[AD^*C]$	$BD^*C\ (y)(4y^*)$
$[AD^*D]$	$BD^*D\ (y)(2y^*)$

NHT =	−9	−7	−5	−3	−1
	2AA*A	4AA*B 2AB*A 4BA*A	6AB*B 8BA*B 6BB*A 2AA*C 2CA*A	6AB*C 4CA*B 2BC*A 6CB*A 2AC*B 4BA*C 16BB*B	4AC*C 2CA*C 4CC*A 2AB*D 2DB*A 14BB*C 10BC*B 14CBB*
SUM =	2	10	24	40	52

Figure D-1.

As an example of the impact of the availability coefficients, recall the general convolutional format for second order fusion as portrayed in Figure D-2: at the stage portrayed below it is poised to generate the terms for $n = -5$, which, from Figure 5-13 are ($[AC^*A]$, BB^*A, CA^*A), (AB^*B, BA^*B) and AA^*C.

+6	+4	+2	0	−2	−4	−6	→

A	B	C	D

Figure D-2.

Table D-2

Available Junctions	Available Antiquirks	Available Junctions	Available Antiquirks
AA*(x)	2d*	CA*(x)	2d*
AB* (2x)	2(d* + u*)	CB*(2x +2y)	4d* +2(d* + u*)
AC* (x)	2u*	CC*(4y +x)	4(d* +u*) + 2u*
AD* (0)	0	CD*(2y)	2u*
BA* (2x)	2d*	DA*(0)	0
BB* (4x +y)	4(d*+u*) +2d*	DB*(y)	2d*
BC* (2x + 2y)	2(d*+ u*) +4u*	DC*(2y)	2d* +2u*
BD* (y)	2u*	DD*(y)	2u*

However, if we take into account the enhancement indicated in the Available Antiquirks columns in Table D-2, as well as the enhanced values of the set A, B, C and D listed previously, the inner products yield the revised six terms (which match the NHT = −5 column in Figure D-1 and reproduced above) as follows:

$AC^*(2u^*)\ d!A = 0$

$AB^*\ (2d^* + 2\ u^*)\ (2d + u)B = 6AB^*B$

$BB^*\ \{4(d^* + u^*) + 2d^*\}\ d!A = 6BB^*A$

$BA^*\ (4d^*)\ (2d + u)B = 8BA^*B$

$CA^*\ (2d^*)\ d!A = 2CA^*A = 2CA^*A$

$AA^*\ (2d^*)\ (d + 2u)\ C = 2AA^*C$

(Note that we have written d!A instead of 3dA in order to indicate that the second order fusion term with three quirks operates as a single point of fusion in this situation.)

The logic of the example used to illustrate how the availability information comes about translates into a formalism as follows: first we note that the enhanced products of first order fusion (in the Available Junctions columns of Figure D-2) are in the inner product form $\alpha_\mu\beta_\nu(rx + sy)$ meaning that (r) FMS with (x) type junctions and (s) FMS with (y) type junctions can be formed in each case. The logic displayed in the example then translates into an expression for an enhanced product with the antiquirks available for fusion in the form

$$\alpha_\mu\beta_\nu\{(r[(\mathrm{pd}^* + \mathrm{qu}^*) - \mathrm{d}^*] + \mathrm{s}[(\mathrm{pd}^* + \mathrm{qu}^*) - u^*]\}. \qquad \text{(D-1)}$$

Here $(\mathrm{pd}^* + \mathrm{qu}^*)$ expresses the composition of the conjugate FMS prior to the first fusion and the two subtractions indicate that a single d^* antiquirk is lost in the formation of each x type junction and a single u^* antiquirk in each y type junction. An equivalent expression is readily obtained in the form

$$\alpha_\mu\beta_\nu\{(r + \mathrm{s})(\mathrm{pd}^* + \mathrm{qu}^*) - (\mathrm{rd}^* + \mathrm{su}^*)\} \qquad \text{(D-2)}$$

which we interpret as subtracting (r) of the (d^*) and (s) of the (u^*) antiquirks from the total of $(\mathrm{r} + \mathrm{s})$ FMS, each with an original $(\mathrm{pd}^* + \mathrm{qu}^*)$ composition of antiquirks. For example, for the case of BB^* we have $\mathrm{r} = 4$, $\mathrm{s} = 1$, $\mathrm{p} = 2$ and $\mathrm{q} = 1$ which computes to

$$(4 + 1)(2\mathrm{d}^* + \mathrm{u}^*) - (4\mathrm{d}^* + \mathrm{u}^*) = 6\mathrm{d}^* + 4\mathrm{u}^* \qquad \text{(D-3)}$$

in agreement with Figure D-2.

To recapitulate, the term $(r+s)$ is the multiplicity associated with the first fusion due to the potential for r, x-type junctions and s, y-type junctions. Under the (conditional) hypothesis that no antiquirks are used up in that process, $(r + s)$ variations of the fused word represented by BB^* would be available for the second fusion, each with two d^* antiquirks and one u^* antiquirk. That accounts for the first term in expression (C.2). In ***reality***, however, antiquirks are lost to junction formation, so a term $(\mathrm{rd}^* + \mathrm{su}^*)$ must be subtracted from the first term to indicate the net availability of antiquirks for the second fusion.

Appendix E

The Beta Switch

Figure E-1 is a knot theoretic demonstration, following the idea [26] that the difference between two diverse intersection situations is

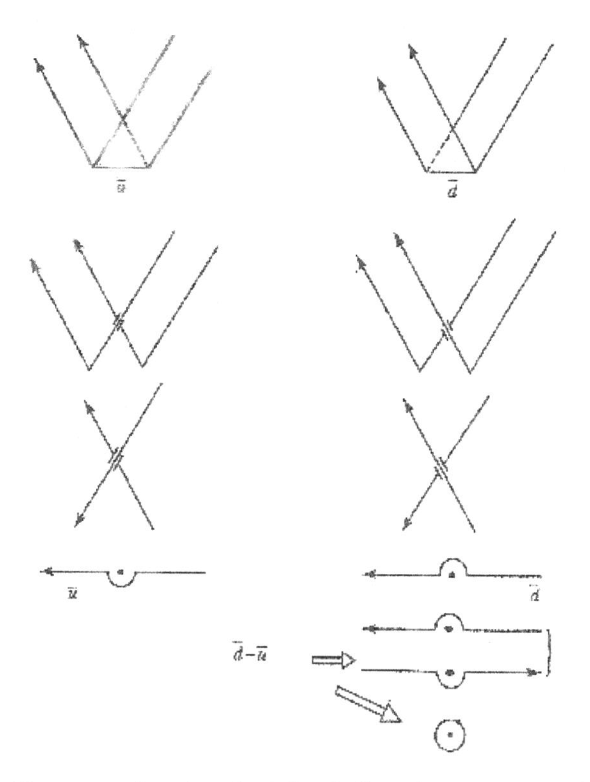

Figure E-1. Demonstration that Anti Quirk d* – Antiqui-Quirk u* equals the unknot.

equivalent to the unknot, in order to demonstrate the validity of the Beta Switch invoked in the Beta decay modeling employed in the preceding. To begin with we show the u* and d* antiquirks on the left and the right, respectively, followed by their respective knot borders. These are then shown unfolded so as to emphasize the two intersection situations, the rightmost segment crossing ***under*** the leftmost segment in the case of u* and ***over*** for d*. The difference, d* − u* is then equivalent to the "unknot" which is nugatory; it can be eliminated, having no effect on the reality of a given situation involving knots.

By way of contrast, Beta decay in the SM is diagrammatically represented in its entirety by the elegant diagram of Figure E-2. The illustrated process looks simple; a ***neutron*** changes into a ***proton*** and a **W** boson, by ***somehow*** changing a ***down quark*** into an ***up quark*** and the **W**, which in turn, then mysteriously changes into an electron and an electron antineutrino; what's missing is "***how***" all that occurs. As far as I know, that has never been explained.

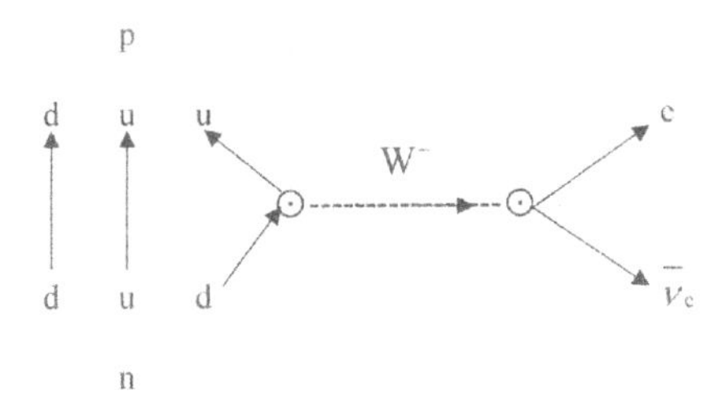

Figure E-2.

Appendix F

The Alternative Model, DNA and Dark Matter Redux

As mentioned in the supplementary, concluding remarks of Chap. 43 of Sec. VII, this Appendix is a result of ruminating upon what many would consider just a superficial resemblance between the components of DNA and the four basic FMSs of our Alternative Model of the elementary particles. The emphasis will be on the "fourness" of each, their two-stranded structures being regarded as simply factual in the particle case and, in the case of DNA, necessary for it to work.

F.1. Summary of DNA (Deoxyribonucleic acid)

By now its basics are common or at least readily available knowledge but I will run through them anyway (according to my non-expert understanding of available expertise [cf. 74]) just so we are on the "same page" so to speak. As we know the famous double helix consists of two potentially parallel strands winding about each other. When (or if) they are uncoiled, the two strands taken together would exhibit a ladder-like structure encompassing what are known as "nucleotides" of which there are four varieties, or more accurately, two sets of two. Each member of one set is composed of ***three*** molecules, a phosphate, a sugar and a base known as a ***Pyridimine***, which has a hexagonal ring-like structure (with a particular atom at each of the corners). Each member of the other set has the same phosphate and sugar and a base known as a ***Purine*** composed of ***two*** ring-like structures, one hexagonal and one pentagonal, linked together.

The phosphate and sugar in each of the four are also linked together to form a segment in the resulting sequence of alternating phosphates and sugars that constitutes each leg of the ladder. The "***rungs***" of the ladder are where the familiar "Genetic Code" resides and it works like this: First of all, the detailed ***chemistry*** of the two Pyridimines (their atomic constituents and the locations thereof) are different and so is the detailed chemistry of the Purines, thus giving us the familiar Pyridimines, ***Thymine*** and ***Cytosine*** (T and C) and the two Purines, ***Adenine*** and ***Guanine*** (A and G), a ***fourfold*** set (and hence the first obvious top-level connection to the basic four FMSs of our Alternative Model!). Figure F-1 shows the four bases, the Phosphate–Sugar segments of the legs of the ladder being of little concern in what follows.

At this point, detailed geometry enters into the picture: each of the four bases is securely linked to a sugar molecule of a segment of the associated ladder leg (T and C by a nitrogen atom of their hexagonal ring and A and G similarly by such an atom of their pentagonal ring) in such a way as to extend ***inward*** relative to the ladder and normal to its legs. As is well known, two young men, Francis **Crick**, a British ex-physicist and James **Watson** a 24-year-old newly-minted Ph.D. from America, won a Nobel prize for uncovering the geometrical structure of DNA. A fascinating account of the race to the Nobel is given in Watson's book, *The Double Helix*, [69]. What the two did was to construct what amounts to a large, laboratory scale 3-D mock-up of the structure on the basis of all kinds of inputs including X-ray photos of DNA samples and chemical experiments conducted in laboratories in a number of countries.

We pick up the story at a point when most of the mistakes and false leads were over and the pair had more or less settled on a two-strand helix, as per the above on the outside of some kind of two-by-two pairing of the four nucleotides. By that time a lot of detailed chemical information was available to the DNA community including some inexplicable statistics on the relative amounts of the four nucleotides; there appeared to be similar amounts of Adenine

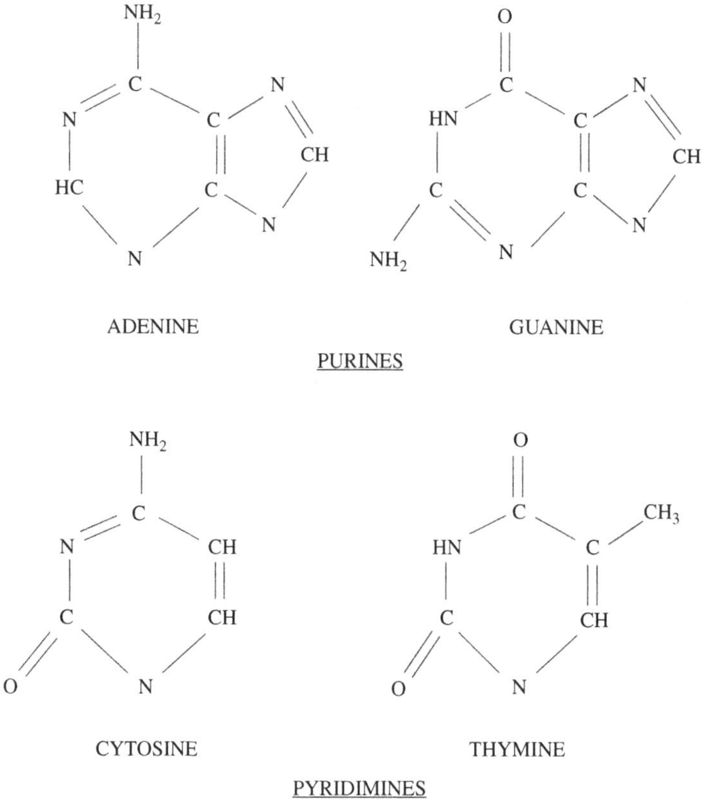

ADENINE GUANINE

PURINES

CYTOSINE THYMINE

PYRIDIMINES

Figure F-1. The four nucleotide bases.

and Thymine as well as of Guanine and Cytosine. Also, Watson and Crick were trying for a smooth structure and, because of the relative sizes, pairing of Purines together and Pyridimenes together made for awkward bulges and twists; the former were too long and the latter too short. To make a too-long story shorter, the choice of pairing G with C and A with T not only made a perfect fit but explained the relative abundance found in chemical experiments.

F.2. DNA/FMS Quaternarity

Another way to think about the two varieties of "rung" (AT) and (GC), on the two-strand DNA ladder is to ***relabel*** them as, respectively,

$$\text{AT} \Leftrightarrow R_A \quad \text{and} \quad GC \Leftrightarrow R_G. \tag{F-1}$$

Also, we note that each of these rung varieties can occur in a directional way (in the direction normal to the legs), say from "left" to "right" — call that (+) — and, conversely, from "right" to "left" — call that (−) — implying that we need a more detailed labeling as, for instance, both

$$R_{A(+)} \quad \text{and} \quad R_{A(-)} \tag{F-2}$$

and, similarly, both

$$R_{G(+)} \quad \text{and} \quad R_{G(-)}, \tag{F-3}$$

again, of course, a fourfold set but expressed in a kind of non-specific way as composed of two ***varieties*** each with two algebraic signs. In fact, if we express the four in a two-dimensional array as

$$\begin{pmatrix} R_{G(-)} & R_{G(+)} \\ R_{A(-)} & R_{A(+)} \end{pmatrix}, \tag{F-4}$$

it is reminiscent (at least to me!) of our fourfold basic FMS array discussed in Sec. II,

$$\begin{pmatrix} \text{B} & \text{C} \\ \text{A} & \text{D} \end{pmatrix}, \tag{F-5}$$

If, that is, the latter is also re-expressed, but here, in terms of both magnitude and sign of ***twist***, for instance as

$$\begin{pmatrix} \text{T}(-) & \text{T}(+) \\ \text{T}^3(-) & \text{T}^3(+) \end{pmatrix}, \tag{F-6}$$

with "T" for twist. In other words, the upper row represents B and C, each with **one** half twist (NHT $= 1$) to the left and right respectively, and similarly, the lower row represents A and D, each with ***three*** half twist, also to the left and the right. The implication is that our basic set of FMS and the set of basic DNA nucleotides have the same ***structural*** representation, still a top-level commonality but no longer just a shared number four!

F.3. Complementarity

So the "obviousity" of the connection mentioned above was right after all! But we cannot leave it at that; for one thing, each structure must undergo a radical schism. In the case of DNA, it is the way cells are replicated in the process of reproduction of the species; the ladder splits (actually, it sort of "unpeels") at that tenuous nucleotide connection, into two single strands studded by ***complementary*** sequences of purines and pyridimines each of which needs to rebuild its partnership by gaining a replacement for its old partner; a pyridimine for each purine and a purine for each pyridimine from a supply of such floating about in the vicinity. Once that happens for each of the two strands: Voila! We then have ***two*** sequences of rungs each identical to the original. In other words, two identical double helices where there was only one before. Amazing! What a clever plan!

We should stress that, since the purines are substantially ***longer*** than the pyridimines, each single strand after the schism has a (seemingly) random, comb-like structure and, as a result, the ***epistemological*** key to the entire process; namely the ***complementarity*** of the two sequences of bristles becomes apparent (to the mind's eye at least!). Actually, the significance of this "self-complementary" nature of the entire DNA structure did not go unnoticed at the time it emerged and is, in fact, noted in Watson's book as, in for instance "— the overwhelming merit of a self-complementary DNA molecule."

So, how about our basic FMS in the Alternative Model; is there a similar situation there? Well, actually we talked about that quite a bit in Sec. II; complementarity **does** occur but in a different way. In the first place, as you may recall, the set of four basic FMS — the A, B, C and D — was separated into two ***complementary*** but overlapping sets of three, A, B and C on the one hand and D, C and B on the other. The development of the taxonomy was then carried out with both sets even though only the first set could provide a match to the Standard Model.[1] And at both stages of taxonomical development — first and second order fusions — we ended up with ***two*** complementary sets; wherever an A appears in the "spelling" of a compound particle in one set, a D appears in the same place in the other set and similarly for B and C; a B in one was matched by a C in the other and a C in one by a B in the other.

Well, that's about it for comparison purposes, at least on a top-level, schematic sort of way. Although the ***geometry*** of either situation is clearly crucially important, as **Crick** and **Watson** discovered in the DNA case and, as we have seen in Sec. II in the case of our model, the domains of the two cases are so disjoint and in such an ***essential*** way, that ***direct*** comparison is precluded; where our Alternative Model applies by definition at the elementary particle level, DNA is pretty much about how biochemistry is involved in reproduction. We need to dig deeper.

F.4. The Depths of the FMS

Figure F-2 is a copy of Figure 2-3 of Sec. II; it shows the very simple geometry of our four basic elementary particles (as compared to the intricate molecular structure of the four nucleotide bases shown in Figure F-1) as well as the simple complement of quirks shared by each figure that illustrate how to distinguish one particle from the other.

The makeup of quirks for each particle in the set can then be summarized by this array:

$$\begin{pmatrix} \mathrm{B} & \mathrm{C} \\ \mathrm{A} & \mathrm{D} \end{pmatrix} \Rightarrow \begin{pmatrix} \mathrm{ddu} & \mathrm{duu} \\ \mathrm{ddd} & \mathrm{uuu} \end{pmatrix}, \tag{F-7}$$

quite succinct enough for our purpose here but it is of interest that we can go one step further. So, recalling how, in Sec. II, the notion

[1]You may recall how the second set was speculated in Sec. VI to constitute a complementary "dark" universe! So, in that regard, read on to the last page.

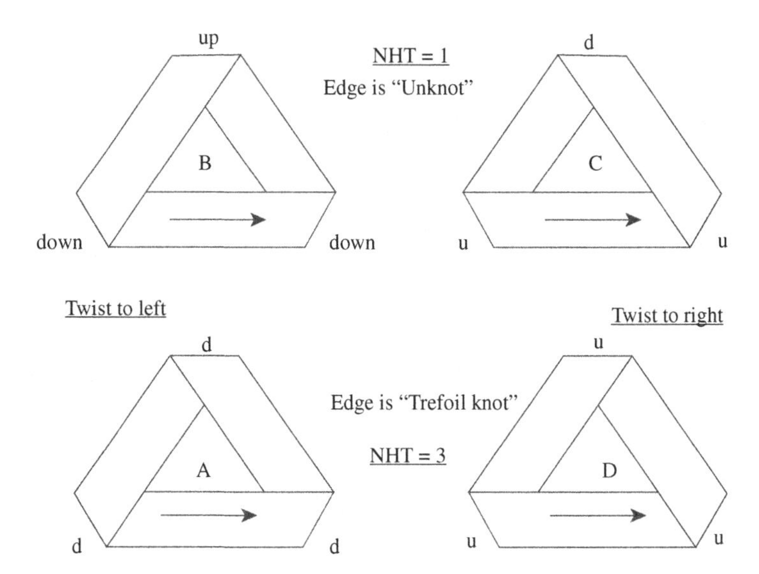

Figure F-2. Copy of basic set of Alternative Model particles.

was introduced that quirks can be viewed as very small steps in time, suppose we now identify an up quirk and a down quirk with time as follows:

$$u \Rightarrow i\tau \quad \text{and} \quad d \Rightarrow -i\tau, \tag{F-8}$$

where

$$i = \sqrt{-1},$$
$$\tau = c\delta t,$$

δt is the size of the step in time and c is the speed of light. This is in a conceptual spacetime where time (but not space) is considered to be an imaginary dimension (we are thinking here about a Minkowski metric plus a Wick rotation). And further, suppose we view the quirks as ***operators*** so that the arithmetic here is ***multiplication*** (rather than addition which, you may recall, we used to identify particle A with the electron, B with the neutron or a neutrino, and C with the Proton).

Then, given the representation of Eq. (F-7) we find that the new representation of our basic four as a 2D array becomes

$$\begin{pmatrix} \mathrm{B} & \mathrm{C} \\ \mathrm{A} & \mathrm{D} \end{pmatrix} \Rightarrow \begin{pmatrix} \mathrm{ddu} & \mathrm{duu} \\ \mathrm{ddd} & \mathrm{uuu} \end{pmatrix} \Rightarrow \begin{pmatrix} -1 & 1 \\ 1 & -1 \end{pmatrix} \tau^3 i \qquad \text{(F-9)}$$

(you do the math!) which represents a small, alternating group with only ***two*** elements namely $\pm i$, thus bringing the representation of our basic set down to bare essentials as an elementary expression of complementarity! Which is as far as we can go; you cannot get much simpler than back and forth — or up and down, etc. — unless you want to stick with only one or the other, i.e. do nothing, neither of which expresses complementarity.

F.5. The Geography of DNA

So, can we do something similar with the nucleotides? On the face of it, it certainly seems doubtful. In a way, the problem is an embarrassment of riches, a rather shopworn phrase but apt here. Consider again Figure F-1 showing the basic chemistry of Purines and Pyridimines; you have to admit that those structures are somewhat more complex than that of the four basic elementary particles! Furthermore, they express a completely different language. So how can we bring that kind of representation down to something that is at least reminiscent of how we talked about the elementary particles?

Well, in that case, the elementary geometry of our particles really comes down to the presence of only ***three*** ***"features"***, namely the three ***quirks***, each with a ***label*** from an alphabet of just ***two*** letters. So, here is the plan: to begin with, we need to reduce the number of features per nucleotide diagram and the way we will do that is to subdivide each diagram into ***regions*** each of which will include a number of atomic constituents, and that will be ***labeled*** such that each diagram is represented by a ***set*** of labels. The idea is to compare the four sets of labels and note the pattern of coincident and noncoincident labeling that results. The trick is to do the subdividing in a judicious manner and, to begin with, it was noted that the pentagonal loops looked like an obvious object of special treatment since they are associated only with the Purines. That being the case, the final pattern of subdivision that emerged after a bit of cogitation is shown in Figure F-3 and the associated pattern of coincidences, indicated by a (+) and non-coincidences is indicated by a (−) is shown in Table F-1. The table was constructed with Adenine as a reference so that all four of its regions were denoted by (+).

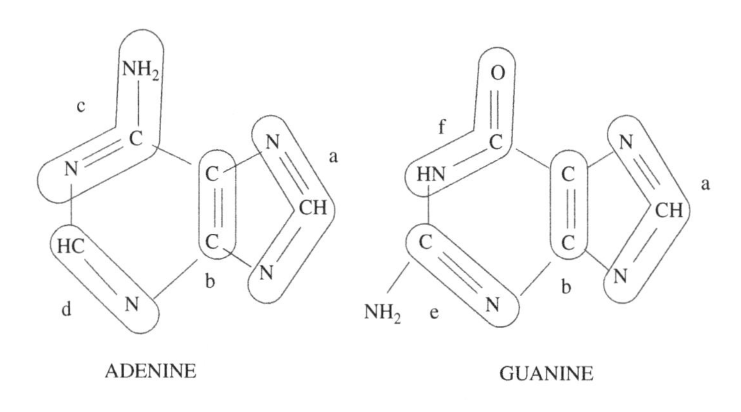

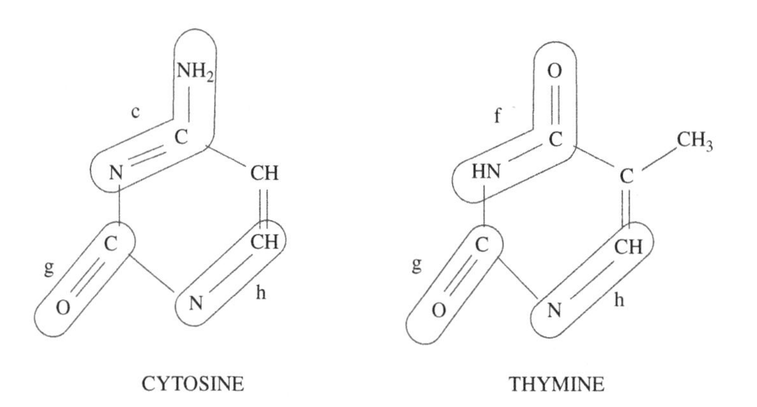

Figure F-3. Nucleotide bases with subdivisions.

Table F-1: Results of labeling comparisons.

	a	*b*	*c*	*d*	*e*	*f*	*g*	*h*
A	+	+	+	+	−	−	−	−
G	+	+	−	−	+	+	−	−
C	−	−	+	−	−	−	+	+
T	−	−	−	−	−	+	+	+

Table F-2: Tabular comparison of two reference selections.

Reference is Adenine				Reference is Thymine			
	a	*b*	*c*	*f*	*g*	*h*	
A	+	+	+	+	+	+	T
G	+	+	−	−	+	+	C
C	−	−	+	+	−	−	G
T	−	−	−	−	−	−	A

As we see, Adenine has four labeled areas with the labels *a*, *b*, *c* and *d*, not all of which apply to all of the four nuclotides. However, the diversity of chemical structure (Of course necessary for the nucleotides to do their job!) has resulted in the need for more labels, a total of eight: *a*, *b*, *c*, *d*, *e*, *f*, *g* and *h*. Nevertheless, a closer look at the table reveals that labels *a*, *b* and *c* suffice (*d*, *e*, *f*, *g* and *h* are superfluous) to do the job; that is, equating + and − to u and d, respectively, then produces a parallel to the labeling of our particles. (Thus the sequence A, G, C and A here corresponds to the sequence D, C, B and A of our particle model.) In other words, we have shown at least one way to validate the resemblance between our particle model and DNA not just superficially but on a detailed level!

F.6. More Geography

Upon closer scrutiny of Figure F-3, however, we see another way to achieve the same result: use ***Thymine*** as a reference. In this case, labels *f*, *g* and *h* suffice and *a*, *b*, *c*, *d* and *e* are superfluous. Table F-2 show comparison of the two cases.

These two tables are left-to-right mirror images but they are ***not complementary***; to demonstrate that we need to show the ***broken*** symmetry in evidence from the point of view of either of the nucleotides as reference, and seen from the associated helical strand after separation as discussed above (in Sec. F.3). We show that in Table F-3 for all four nucleotide combinations using Adenine as a reference but it becomes apparent if any of the other three are used in that way.

Table F-3: Complementary nucleotide tables (Reference is Adenine).

	a	b	c	c	b	a	
A	+	+	+	−	−	−	T
G	+	+	−	+	−	−	C
C	−	−	+	−	+	+	G
T	−	−	−	+	+	+	A

The table on the right is drawn simply by rotating the table on the left 180° about an axis normal to the paper just to make the complementarity obvious at a glance but a closer look shows that the sequence $abc = + + -$ for Guanine is matched by its complementary sequence $abc = - - +$ for Cytosine and so forth for the other three nucleotides A, C and T and their complements T, G and A. Furthermore, it is straightforward to show the same relationships obtain using T, G or C as references. But, notice: no matter what reference is used, the complementary pairings are A with T and G with C! Which, as Crick and Watson noted, explains the statistical findings they had puzzled over. It looks like we are getting somewhere!

So now: back to the question posed above as to whether we can reduce the representation down any further, as we did for the particles. Well, suppose we start with the representation $\begin{pmatrix} G & C \\ A & T \end{pmatrix}$ for the nucleotides with, as in the above, Thymine as the reference. (Again, we can use any of the other three nucleotides as the reference.)

Then, using Table F-2, this translates to $\begin{pmatrix} +-- & -++ \\ --- & +++ \end{pmatrix}$ which is essentially the quirk representation for the particles, except for the latter's use of imaginaries. (Again, we can use any of the other three nucleotides as the reference.) Although I see no way to justify using imaginaries in the case of the nucleotides we do not need them; there is no less reason to treat the algebraic signs multiplicatively here than there was before and when we do, we arrive at $\begin{pmatrix} + & - \\ - & + \end{pmatrix}$, again, a two-state representation of complementarity as for the particle case.

Well, that ought to be enough — or close anyway. The only thing we haven't talked about here is the requirement for symmetry breaking. Complementarity comes in many forms; ours requires two sequences (in our case, the bare minimum of two items in each) in a mirror relationship that has to be broken to produce it. Well, we saw that too, in both cases. In the case of the elementary particles of the Alternative Model it was due to a necessary choice for a direction of traverse of the FMS which ended up being responsible for the two complementary sets and, ultimately, the (putative) existence of the two complementary universes. And in the case of genetics, it's the need to have both a Purine and a Pyridimene base, with their different lengths, (lightly) bonded together so as to make up each "codon" of the genetic Code in order that we end up with two complementary codes after the "unzipping" and, in the end of the process, two copies of the original DNA molecule.

F.7. Reflections

Well, that's it; I believe we've done it! You know when I started out to put this Appendix together I really didn't know how it would turn out so the outcome is most gratifying; it's not every day that one witnesses a fundamental resemblance emerging between the structures of two such diverse phenomena as DNA and the elementary particles of Physics! Still, DNA behavior, that is its essential replication of the complementarity inexorably embedded in our basic particle set as recalled above, is, on some level, rather inexplicable. How can that be when the individual scales are so many orders of magnitude apart? Can it be that what transpires at the larger scale is, ***somehow***, an inevitable outcome of the structure on the smaller scale? Which in turn evokes the similar question: can it be that the association, as in Sec. VI, of ***Dark Matter*** (whose effects we witness on a cosmological scale) with the subset D, C and B of our basic quartet of elementary particles is also an inevitability rather than just speculation? If you recall, I postulated that association in the first place because of the complementary relationship of that subset of elementary particles (which seemed to have no existential purpose) with its partner the subset A, B and C that we linked to the Standard Model.

These seem like just rhetorical questions but let's pause and take stock here: do you see what I see? What I see is a very similar basic structure, pattern, scheme, etc. in the characteristics of both of those two very diverse ***systems***. First there's the basic "fourness"; four ***objects*** that constitute the basis for each system: a set of four ***FMS*** for the basic particles and a set of four ***Nucleotides*** for DNA. Then in each case, the basic set is divided into two ***complementary*** sets. In the case of the particles one set is the basis for our universe and the other is (postulated to be) the basis for the "Dark" universe. In the case of DNA, Table F-3 shows two such sets and suggests the match-ups (A with T and G with C) that form the "rungs" of the ladder, and which lead to genetic mitosis and, ultimately the perpetuation of a species. Each object is in turn described in terms of a set of three ***features***, the ***quirks*** for the particles and the molecular "***regions***" for the Nucleotides. And, finally, each set of features is represented by a "***word***" whose letters comprise a ***binary alphabet***. In summary: four object, three features and two letters: ***four, three, two***. Plus the division into two ***complementary*** sets.

One must confront the unavoidable and we seem to be faced here with not just an interesting set of circumstances; it's beginning to look like a ***Principle***! And perhaps of greater generality than just our two examples treated here. For a discussion of that, please turn back to Sec. VII, specifically the concluding remarks of Chap. 43. But to return to the questions posed above: it would appear that it's not a mysterious influence at the smaller scale that causes the behavior at the larger scale; it's just that all three effects are dur to a common requirement or set of requirements; a ***principle*** whose salient feature is ***complementarity***.

So, as far as this book is concerned, having tied everything up, at least to my own satisfaction, I shall end it right now before I come up with more things to talk about and reveal too much of my ignorance. Let me know what you think; about anything, in the book or not. I don't do Twitter or Facebook but this should work:

javrin@aol.com

References

[1] A. Vilenkin, *Many Worlds in One, The Search for Other Universes* (Hill and Wang, 2006)

[2] A. A. Filippov, *The Versatile Soliton* (Birkhäuser, 2000)

[3] E. T. Bell, The *Development of Mathematics*, 1940. (Courier Dover Publications, 1945).

[4] W. H. Thompson, *Trans. Roy. Soc. Edinburgh* **25** (1869) 217–220.

[5] E. Witten, Reflections on the fate of spacetime, *Phys. Today* (April, 1996).

[6] U. Enz, *Phys. Rev.* **131** (3) (1953).

[7] T. H. R. Skryme, *Proc. Roy. Soc. A* **247** (1958) 260; *ibid.* **252** (1959) 236.

[8] R. L. Ricca and M. A. Berger, Topological ideas and fluid mechanics, *Phys. Today* (December, 1996) 28–61.

[9] L. Fadeev and A. J. Niemi, Stable, knotlike structures in classical field theory, *Nature* **387** (197) 58–621.

[10] J. S. Avrin, *J. Knot Theory Remifications* **14** (2005) 131–176.

[11] J. S. Avrin, *J. Knot Theory Remifications* **17** (2008) 835–876.

[12] J. S. Avrin, *J. Knot Theory Remifications* **21** (2012) 1250004.

[13] J. S. Avrin, *J. Knot Theory Remifications* **20** (2011) 1723–1739.

[14] J. S. Avrin, *Symmetry* **4** (2012) 39–115.

[15] S. Sternberg, *Group Theory and Physics* (Cambridge Univ. Press, 1994).

[16] Webster's New World Dictionary of the American Language.

[17] C. Itzykson and J.-B. Zuber, *Quantum Field Theory* (McGraw-Hill, 1980).

[18] Encyclopaedia Britannica, '*Ptolemy (Claudius Ptolemaius)*' (William Bention, 1964).

[19] Wikepedia, the free encyclopedia, "*Nicolaus Copernicus*".

[20] McGraw-Hill Encyclopedia of Physics, 2nd edn. (McGraw-Hill, 1992).

[21] E. Fermi and C. N. Yang, *Phys. Rev.* **76** (12) (1949) 1739–1743.

[22] Y. Nambu, *Quarks, Frontiers in Elementary Particle Physics* (World Scientific, 1985).

[23] S. Sakata, *Prog. Theor. Phys.* **16** (1956) 686.

[24] S. Bart, *Experiments in Topology* (Thomas Y. Crowell Co., 1964).

[25] Ch. J. Isham, *Modern Differential Geometry for Physicists*, 2nd edn. (World Scientific, 1999).

[26] L. H. Kauffman, *Knots and Physics*, 3rd edn. (World Scientific, 2001), p. 38.

[27] E. Flapan, *When Topology Meets Chemistry* (Cambridge Univ. Press, 2000).

[28] A. Sossinsky, *Knots; Mathematics with a Twist* (Harvard Univ. Press, 2002).

[29] L. O'Raifeartaigh, *The Dawning of Gauge Theory* (Princeton Univ. Press, 1997).
[30] G. Kuperberg, Non-involutory Hopf algebras and 3-manifold invariants, arXiv:q-alg/9712047v1
[31] E. Cartan, *The Theory of Spinors* (Dover, 1966).
[32] M. Carmeli and S. Malin, *An Introduction to the Theory of Spinors* (World Scientific, 2000).
[33] Goudsmit and Uhlenbeck, McGraw-Hill Encyclopedia of Physics, 2nd edn. (McGraw-Hill, 1992).
[34] S. L. Altmann, *Rotations, Quaternions and Double Groups* (Dover, 1985).
[35] H. S. Black, *Modulation Theory* (D. Van Nostrand Company, 1953).
[36] W. H. Rolnick, *The Fundamental Particles and Their Interactions* (Addison Wesley. 1994).
[37] C. Lanczos, *The Variational Principles of Mechanics*, 4th edn. (Dover, 1970).
[38] W. A. Perkins, *Int. J. Mod. Phys. A* **16** (2001) 919–921.
[39] K. Gottfried and V. Weisskopf, *Concepts of Particle Physics*, Vol. 1 (Oxford, 1984).
[40] G. D. Coughlan and J. E. Dodd, *The Ideas of Particle Physics, An Introduction for Scientists*, 2nd edn. (Cambridge Univ. Press, 1991).
[41] J. Baez, http://math.ucr.edu?home/baez/
[42] E. Kreysig, *Differential Geometry* (Dover, 1991).
[43] S. Weinberg, *Gravitation and Cosmology, Principles and Applications of the General Theory of relativity* (John Wiley & Sons, 1972).
[44] E. Cartan, *On Manifolds with an Affine Connection and the Theory of General Relativity*, trans. A. Magnon and A. Ashtekar (Bibliopolis, 1986).
[45] A. Einstein, *On the quantum theorem of Sommerfeld and Epstein*, translation in *The Collected Papers of Albert Einstein*, Vol. 6, ed. A. Engel (Princeton Univ. Press, 1997).
[46] D. Stone, Einstein's unknown insight and the problem of quantizing chaos, *Physics Today* (August 2005).
[47] M. L. Irons, The curvature and geodesics of the torus, http://www.rdrop.com/-half/math/torus/index.xhtml, 17, November 2005.
[48] B. Greene, *The Elegant Universe, Superstrings, Hidden Dimensions, and the Quest for th Ultimate Theory* (W. W. Norton & Co., 1999).
[49] F. Mandl and G. Shaw, *Quantum Field Theory,* Revised edn. (John Wiley Sons, 1993).
[50] J. A. Wheeler, *Geons, Black Holes & Quantun Foam* (W. W. Norton & Co., 2000).
[51] I. Ciufolini and J. A. Wheeler, *Gravitation and Inertia* (Princeton Univ. Press, 1995).
[52] M. Kaku, *Quantum Field Theory, A Modern Introduction* (Oxford Univ. Press, 1993).
[53] P. D. B. Collins, A. D. Martin and E. J. Squires, *Particle Physics and Cosmology* (John Wiley & Sons, 1989).
[54] J. Lykken and M. Spiropulu, Supersymmetry and the crisis in physics *Scientific Amer.*, August 24–September 7 (2014).

[55] N. L. Libeskind, Dwarf galaxies and the dark web, *Scientific Amer.*, March (2014).
[56] A. Linde, The self-reproducing universe, *Scientific Amer. Sp. Issue On The Cosmic Life Cycle of the Universe.*
[57] E. Witten, *Comm. Math. Phys.* **117** (1988) 353–399.
[58] E. Witten, Quantum field theory and the Jones polynomial, in the *Proc. of the 1988 IAMP Congress, Swansea Comm. Math. Phys.* **121** (1989).
[59] M. Atiyah, *The Geometry and Physics of Knots* (Cambridge Univ. Press, 1993).
[60] G. Musser, *The Complete Idiots Guide to String Theory* (Alpha Books, 2008).
[61] S. G. Gubster, *The Little Book of String Theory* (Princeton Univ. Press, 2010).
[62] M. Rees, *Just Six Numbers, The Deep Forces that Shape the World* (Basic Books, 2000).
[63] J. D. Barrow and F. J. Tipler, *The Anthropic Cosmological Principle* (Oxford Univ. Press, 1986).
[64] L. H. Kauffman, *Constructivist Foundations* **4** (2009) 121–137.
[65] L. H. Kauffman, Private Communication.
[66] Ed. A. Ashtekar, *100 Years of Relativity, Space-Time Structure: Einstein and Beyond* (World Scientific, 2006).
[67] L. Smolin, *The Trouble with Physics, The Rise of String Theory, The Fall of a Science, and What Comes Next* (Houghton Mifflin, 2006).
[68] E. Witten, *Philos. Trans. R. Soc. London* **329** (1989) 349–357.
[69] J. D. Watson, *The Double Helix, A Personal Account of the Discovery of the Structure of DNA* (Simon & Schuster, 1968).
[70] T. Curtright, and D. Fairlie, *Phys. Rev. D* **58** (1998) 025002.
[71] M. I. Skolnick, *Introduction to Radar Systems* (McGraw-Hill, 1962).
[72] D. E. Neuenschwander, *Emmy Noether's Wonderful Theorem* (The Johns Hopkins Univ. Press, 2011).
[73] C. Mulcahy and D. Richards, Let the games continue, *Sci. Amer.* **211** 4, October 2014.
[74] R. A. Wallace, *Biology, The World of Life* (Goodyear Publishing Co., 1975).

Index

SERIES ON KNOTS AND EVERYTHING

Editor-in-charge: Louis H. Kauffman *(Univ. of Illinois, Chicago)*

The Series on Knots and Everything: is a book series polarized around the theory of knots. Volume 1 in the series is Louis H Kauffman's Knots and Physics.

One purpose of this series is to continue the exploration of many of the themes indicated in Volume 1. These themes reach out beyond knot theory into physics, mathematics, logic, linguistics, philosophy, biology and practical experience. All of these outreaches have relations with knot theory when knot theory is regarded as a pivot or meeting place for apparently separate ideas. Knots act as such a pivotal place. We do not fully understand why this is so. The series represents stages in the exploration of this nexus.

Details of the titles in this series to date give a picture of the enterprise.

*Published**:

Vol. 1: Knots and Physics (3rd Edition)
by L. H. Kauffman

Vol. 2: How Surfaces Intersect in Space — An Introduction to Topology (2nd Edition)
by J. S. Carter

Vol. 3: Quantum Topology
edited by L. H. Kauffman & R. A. Baadhio

Vol. 4: Gauge Fields, Knots and Gravity
by J. Baez & J. P. Muniain

Vol. 5: Gems, Computers and Attractors for 3-Manifolds
by S. Lins

Vol. 6: Knots and Applications
edited by L. H. Kauffman

Vol. 7: Random Knotting and Linking
edited by K. C. Millett & D. W. Sumners

Vol. 8: Symmetric Bends: How to Join Two Lengths of Cord
by R. E. Miles

Vol. 9: Combinatorial Physics
by T. Bastin & C. W. Kilmister

Vol. 10: Nonstandard Logics and Nonstandard Metrics in Physics
by W. M. Honig

Vol. 11: History and Science of Knots
edited by J. C. Turner & P. van de Griend

*The complete list of the published volumes in the series can also be found at
http://www.worldscibooks.com/series/skae_series.shtml

Vol. 12: Relativistic Reality: A Modern View
edited by J. D. Edmonds, Jr.

Vol. 13: Entropic Spacetime Theory
by J. Armel

Vol. 14: Diamond — A Paradox Logic
by N. S. Hellerstein

Vol. 15: Lectures at KNOTS '96
by S. Suzuki

Vol. 16: Delta — A Paradox Logic
by N. S. Hellerstein

Vol. 17: Hypercomplex Iterations — Distance Estimation and Higher Dimensional Fractals
by Y. Dang, L. H. Kauffman & D. Sandin

Vol. 18: The Self-Evolving Cosmos: A Phenomenological Approach to Nature's Unity-in-Diversity
by S. M. Rosen

Vol. 19: Ideal Knots
by A. Stasiak, V. Katritch & L. H. Kauffman

Vol. 20: The Mystery of Knots — Computer Programming for Knot Tabulation
by C. N. Aneziris

Vol. 21: LINKNOT: Knot Theory by Computer
by S. Jablan & R. Sazdanovic

Vol. 22: The Mathematics of Harmony — From Euclid to Contemporary Mathematics and Computer Science
by A. Stakhov (assisted by S. Olsen)

Vol. 23: Diamond: A Paradox Logic (2nd Edition)
by N. S. Hellerstein

Vol. 24: Knots in HELLAS '98 — Proceedings of the International Conference on Knot Theory and Its Ramifications
edited by C. McA Gordon, V. F. R. Jones, L. Kauffman, S. Lambropoulou & J. H. Przytycki

Vol. 25: Connections — The Geometric Bridge between Art and Science (2nd Edition)
by J. Kappraff

Vol. 26: Functorial Knot Theory — Categories of Tangles, Coherence, Categorical Deformations, and Topological Invariants
by David N. Yetter

Vol. 27: Bit-String Physics: A Finite and Discrete Approach to Natural Philosophy
by H. Pierre Noyes; edited by J. C. van den Berg

Vol. 28: Beyond Measure: A Guided Tour Through Nature, Myth, and Number
by J. Kappraff

Vol. 29: Quantum Invariants — A Study of Knots, 3-Manifolds, and Their Sets
by T. Ohtsuki

Vol. 30: Symmetry, Ornament and Modularity
by S. V. Jablan

Vol. 31: Mindsteps to the Cosmos
by G. S. Hawkins

Vol. 32: Algebraic Invariants of Links
by J. A. Hillman

Vol. 33: Energy of Knots and Conformal Geometry
by J. O'Hara

Vol. 34: Woods Hole Mathematics — Perspectives in Mathematics and Physics
edited by N. Tongring & R. C. Penner

Vol. 35: BIOS — A Study of Creation
by H. Sabelli

Vol. 36: Physical and Numerical Models in Knot Theory
edited by J. A. Calvo et al.

Vol. 37: Geometry, Language, and Strategy
by G. H. Thomas

Vol. 38: Current Developments in Mathematical Biology
edited by K. Mahdavi, R. Culshaw & J. Boucher

Vol. 39: Topological Library
Part 1: Cobordisms and Their Applications
edited by S. P. Novikov & I. A. Taimanov

Vol. 40: Intelligence of Low Dimensional Topology 2006
edited by J. Scott Carter et al.

Vol. 41: Zero to Infinity: The Fountations of Physics
by P. Rowlands

Vol. 42: The Origin of Discrete Particles
by T. Bastin & C. Kilmister

Vol. 43: The Holographic Anthropic Multiverse
by R. L. Amoroso & E. A. Ranscher

Vol. 44: Topological Library
Part 2: Characteristic Classes and Smooth Structures on Manifolds
edited by S. P. Novikov & I. A. Taimanov

Vol. 45: Orbiting the Moons of Pluto
Complex Solutions to the Einstein, Maxwell, Schrödinger and Dirac Equations
by E. A. Rauscher & R. L. Amoroso

Vol. 46: Introductory Lectures on Knot Theory
edited by L. H. Kauffman, S. Lambropoulou, S. Jablan & J. H. Przytycki

Vol. 47: Introduction to the Anisotropic Geometrodynamics
by S. Siparov

Vol. 48: An Excursion in Diagrammatic Algebra: Turning a Sphere from Red to Blue
by J. S. Carter

Vol. 49: Hopf Algebras
by D. E. Radford

Vol. 50: Topological Library
Part 3: Spectral Sequences in Topology
edited by S. P. Novikov & I. A. Taimanov

Vol. 51: Virtual Knots: The State of the Art
by V. O. Manturov & D. P. Ilyutko

Vol. 52: Algebraic Invariants of Links (2nd Edition)
by J. Hillman

Vol. 53: Knots and Physics (4th Edition)
by L. H. Kauffman

Vol. 54: Scientific Essays in Honor of H Pierre Noyes on the Occasion of His 90th Birthday
edited by J. C. Amson & L. H. Kauffman

Vol. 55: Knots, Braids and Möbius Strips
by J. Avrin

Vol. 56: New Ideas in Low Dimensional Topology
edited by L. H. Kauffman & V. O. Manturov

Printed in the United States
By Bookmasters